Urbanising Tanzania

ISSUES, INITIATIVES AND PRIORITIES

Edited by

Suleiman Ngware and J. M. Lusugga Kironde

DUP (1996) LTD.

Published in Tanzania by:

DUP (1996) LTD
University of Dar es Salaam
Uvumbuzi Road
P.O. Box 35182
DAR ES SALAAM

This book has been published with the financial support of Sida

Cover Photo: Courtesy of P. Mgimba

Copy editing, typesetting, and book design: Bernhard J. Sanyagi

ISBN 9976 60 325 8

CONTENTS

LIST OF TABLES AND FIGURES

ACKNOWLEDGEMENTS

Urbanisation is a burning issue in Tanzania not only because of its social and economic significance, but also because of its high rate of growth and lack of effective mechanism to manage its outcomes - such as the pressure on urban services and infrastructure. Between 1980 and 1997, the urban population grew at a rate of 10% per annum. Both the 1978 and 1988 national census figures show that a bigger share of urbanization is a result of rural-urban migration. It is anticipated that 17.5 million people or about 36.3 per cent of the total population in Tanzania will be living in urban areas by the year 2001. Dar es Salaam alone is said to receive between 100,000 to 300,000 people a year. Most of these migrants live in un-serviced areas of towns with environmental conditions which threaten life and health. After the implementation of the structural adjustment programmes, liberalization and the removal of subsidised services in urban areas, poverty is becoming more pronounced.

This volume contains revised versions of papers which were presented at a national workshop held to "Establish a Knowledge Base on Urbanization in Tanzania" in September 1998. The workshop was jointly organised by Swedish Embassy in Dar es Salaam, The Institute of Development Studies of the University of Dar es salaam, and University College of Lands and Architectural Studies (UCLAS) fomerly Ardhi Institute.

In the lengthy and excruciating process leading to the publication of this book, we have been fortunate to receive the committed support of many individuals and institutions. Mr. Thomas Anderson of the Swedish Embassy in Tanzania, the Director of the Institute of Development Studies and the Principal of University College of Lands and Architectural Studies helped in the conception of the whole project and then did whatever was asked of them to keep the project going. Financial support for the book (and the workshop that preceded it) was obtained from Sida through the Swedish Embassy in Tanzania.

The production of this book required the cooperation and painstaking coordination between many institutions and individuals as well as authors and reviewers of the selected papers presented at the national workshop. Special thanks must be given to colleagues at Sida, the Institute of Development Studies, and the University College of Lands and

Architectural Studies (UCLAS). We are also grateful to Ms. Masandeko and Ms. Belinda for their careful typing and re-typing of the original and revised versions of the manuscripts as well as providing invaluable administrative support. Special thanks go to Honourable G. Cheyo, (MP) Minister of Lands and Human Settlements who officiated at the workshop. The Vice-Chancellor of the University of Dar es Salaam, Professor Matthew Luhanga, graced us not only with chairing of the workshop but extended constant moral and material support to ensure that the book comes out on time. For the keen language checking of original manuscripts, we are indebted to Dr. C. Rubagumya, of the Department of Foreign Languages and Linguistics.

After the revisions and re-typing were completed, Bernhard Sanyagi of DUP (1996) LTD., copy-edited the manuscripts and designed the book and page lay-out. We also wish to thank him for his time, patience and advise during many of our consultative meetings.

Lastly, we would like to reiterate our gratitude to the people, and the government of Sweden and to Sida as well as participants from Sweden and Tanzania, who provided the *genius loci* for fruitful work and rich individual experiences. Needless to say, while we are willing to take responsibility for any errors, omissions or short falls which may appear in this book, these collegues and friends should share the credit for what value this work might bring.

Editors, Dar es Salaam, 1999

ABOUT THE CONTRIBUTORS

SULEIMAN NGWARE is currently an Associate Professor with the Institute Development Studies, University of Dar e s Salaam. He researches, teaches and writes on issues of governance, local self-governance and urban development. He is the author and co-editor of numerous publications on the subject.

J. M. LUSUGGA KIRONDE is an Associate Professor with the University College of Lands and Architectural Studies (UCLAS), University of Dar es Salaam - previously the Ardhi Institute. A graduate of Nairobi University in 1995 he was at the time of publishing this book the Registrar at the same College, with numerous publications and consultancies in the field of urbanisation and urban management.

ADOLFO MASCARENHAS - BA (Hon)Lon; MA, Ph.D, UCLA, Dip Ed. (EA) - was recruited by the Government of Tanganyika, as the first cadet Town Planner in the Ministry of Local Government & Administration upon graduating in 1961 from the University of London. He resigned after a few months of service but kept up his interest in the City of Dar es Salaam and subsequently wrote his Masters Thesis for the University of California, Los Angeles. He has specialized in urbanization and also acted as an advisor/ trainer for two of the early population censuses of Tanzania. He has had the opportunity to travel and work in most of the capitals of East and Southern Africa including a five day visit to the historic city of Timbuktu.

FRED S. LERISE is a Senior Lecturer in the Department of Urban and Rural Planning at the University College of Lands and Architectural Studies (UCLAS), formerly called the Aridhi Institute. Trained as an urban and rural planner, he teaches and researches in issues related to the management of urban and rural settlements and general aspects of land use planning. He has accumulated a considerable practical experience through consultancies in the same field.

JOSEPH D. KESSY is a consultant in property valuation and the Chairman of Tabata Development Fund, a community based organisation at Tabata, Dar es Salaam.

ROSE SHAYO is a Tanzanian academic and a consultant on gender training, at the Institute of Development Studies, University of Dar es Salaam, where she has taught for many years. She has published several articles on political, economic as well as on urban development in Tanzania. She has co-edited a book on *Gender and Agrarian Change in Tanzania with a Kenyan Case Study*. Her research interests include impacts of SAP on public sector employment, democratisation etc.

ABBREVIATIONS

CBD	Central Business District
CBO	Community Based Organisation
CIP	Community Infrastructure Project
CTI	Chamber of Tanzanian Industries
DANIDA	Danish International Development Agency
DAWASA	Dar es Salaam Water and Sewrage Authority
DCC	Dar es Salaam City Council (Commission)
DMT	Dar es Salaam Motor Transport
IDS	Institute of Development Studies
ILO	International Labour Organisation
LDCs	Least Developed Countries
MAPET	Manual Pit Emptying System
MUWADA	Dar es Salaam Organisation of Daladala Operators
MLHDS	Ministry of Lands and Human Settlements Development
MRA&LG	Ministry of Reginal Administration and Local Governments
NGO	Non Governmental Organisation
NHC	National Housing Corporation
NIGP	National Income Generation Programme
NORAD	Norwegian Agency for International Development
NUMEIST	National Urban Mobility, Employment and Income Survey of Tanzania
NUWA	National Urban Water Authority
RHF	Revolving Housing Fund
SDP	Sustainable Dar es Salaam Project
Sida	Swedish International Development Agency
SSA	Sub-Sahara(n) Africa
TAMWA	Tanzania Media Women Association
TANESCO	Tanzania National Electric Supply Company
TDF	Tabata Development Fund
TGNP	Tanzania Gender Network Group
TTCL	Tanzania Telecoms Corporation Limited
UASU	Urban Authorities Support Unit
UCLAS	University College of Lands and Architectural Studies
UDA	Usafiri Dar es Salaam (*see* also DMT)
UNCH	United Nations Centre for Human Settlements
UNDP	United Nations Development Programme
URT	United Republic of Tanzania

1
INTRODUCTION

J.M. Lusugga Kironde[1] *and Suleiman Ngware*[2]

There is considerable concern about the rapid urbanisation taking place in Third World countries particularly in Africa. This concern was expressed by the Second United Nations Conference on Human Settlements (Habitat II) which took place in Istanbul, Turkey, in June 1996 and which was aptly called "The City Summit." The subsequent publication: *An Urbanising World: A Global Report on Human Settlements* presented the conditions in urban areas in various regions of the globe, and what is being done to address the problems posed by rapid urbanisation. In most countries of the Third World, governments are poorly equipped in terms of both resources, and appropriate policy initiatives to face this challenge of rapid urbanisation.

There is however, a general agreement that the problems of rapidly growing urban communities must be dealt with concern aimed at dealing with problems of urbanisation. This is a change and a challenge from the attitudes of the past decades where urban communities were considered to be an evil to be fought. Even in the light of high rates of migration of people from rural to urban areas, and the positive relation between urbanisation and economic and social development, governments conceived policies that were anti-urbanisation, such as those aimed at repatriating people back to, or at least preventing them from leaving the rural areas, and doing little or nothing about urban areas. Rural people too regarded urban people with apprehension, and many early migrants to urban areas considered themselves temporary sojourners who would eventually go back to the rural areas.

---

[1] Associate Professor, University College of Lands and Architectural Studies, UDSM
[2] Associate Professor, Institute of Development Studies, UDSM

The situation is changing rapidly and generations of permanent urban residents are being created. Nevertheless, whether residents of urban areas are to stay there for good, or whether they are there on a temporary basis, the problems caused by rapid urbanisation are urgent and can no longer be ignored. They include those related to creating employment and income generating opportunities for the growing populations; providing land, shelter, and physical and social infrastructure; managing the environment and city growth; feeding and transporting the population; putting in place appropriate institutions and mastering resources required to govern the city; growing urban poverty, social tensions and increasing insecurity; a poor information base; and effects of globalisation.

Many of these problems are currently being experienced. While the will to deal with them is there, lack of resources, and inappropriate approaches have limited the impact of interventions from the public authorities. Increasingly, community based organisations are playing positive roles in various urban areas where society is also increasingly demanding better services, as well as forms of governance that emanate from and are responsive to local needs.

The limited impact of public authority action has led to gross inadequacy in the provision of goods and services required for the growing urban areas, and to the development of the informal city. A lot of action is undertaken spontaneously by city residents, many times contrary to the wishes, or plans of public authorities. This has resulted into informal settlements, informal economic activities, unregulated and unprovided for urban development; a breakdown of law and order, slums, and a poor urban environment. There is a convergence of both rich and poor urban residents living in the unserviced city.

It is clear that the future calls for a positive attitude into understanding and managing rapid urbanisation and development. Yet there is limited research that can inform policy makers into the current processes of urbanisation and the resulting popular action towards it. Current approaches are still very much rooted into what Okpala (1987) has termed "received concepts and theories" many of them borrowed from developed countries and characterised by centralisation of powers, unrealistic standards, bureaucratic processes and the non incorporation of popular and community action. There is thus need to carry out research into alternative and appropriate approaches towards this phenomenon of urbanisation. This is more so since, besides rapid urbanisation, Third World countries have been subjected to globalisation and to macro-economic reforms aimed at reducing the direct role of government, and at enhancing the role of the civil society and the private sector. The political climate too has changed with the development of multiparty democracy as opposed to the past where pluralistic democracy was restricted. Various strands of policy reflect the growing concern with the urbanisation phenomenon in the country. For example, both the New Land Policy (Tanzania 1995), and the Land Act of 1998 which grew out of this policy have addressed themselves positively to the problems of access to and managing of urban land, although the continued concentration of powers within the Ministry of Lands and Human Settlements has been criticised.

Proposed by the Ministry of Lands and Human Settlements, an Urban Development Policy has been mooted and deliberated, and is currently being put together. So is the National Human Settlements Policy. Local Government Reform, aimed at creating strong and democratic local authorities is under implementation. The City of Dar es Salaam with an estimated population of 3 million is being restructured into three municipalities, which, it is hoped, will be more responsive to local needs, and be more accountable to the local population. Taking cue from the Sustainable Dar-es-Salaam Project, several municipalities have carried out environmental profiles for their urban areas, have conducted city consultations, and have prioritised their needs. An important element of this "Environmental Planning and Management" approach is the incorporation of stakeholder in determining and executing actions. Many urban authorities no longer ignore community action, and the government is encouraging more interaction between the authorities and the general population.

Academic institutions such as the University of Dar es Salaam are also getting more and more interested in problems of urbanisation not only through reshaping their curricula, but also through conducting relevant research, and setting up new institutions. Centres for environmental and for urban research may be operative in the not so distant future at the University of Dar es Salaam.

The international community is showing increasing interest in urban problems, after decades of concentrating resources on rural areas. Such interest is manifested in the support extended to CBOs, and civic groups and resources set aside for research and information dissemination.

To a large extent, the chapters contained in this book are aimed at contributing to the understanding of, and the growing debate on, urbanisation in Tanzania. Originally, the papers on which these chapters are based were presented at a Sida sponsored two-day workshop (2-3 September 1998) whose aim was to establish a knowledge base on the urbanisation process in Tanzania.

In Chapter 2, Ngware reviews the status of urban research and policies in Tanzania arguing that the rural biased policies of the 1960s and 1970s must now be reoriented towards the urban areas. He is apprehensive that if policy makers and researchers do not take urgent and sustainable measures to address the fundamental issues regarding urban governance, poverty and decay, the country will not achieve sustainable development and will be faced with unnecessary social costs, and the agony of political strife and social havoc. The need to shift from state-centric, to a multi-faceted, multi-sectoral and multi-dimensional paradigm in which the various stakeholders, actors and partners are active players is stressed. Moreover, Ngware complains that hitherto, focus has been on the largest urban areas, ignoring the small and medium sized urban centres which are now an important aspect of the national socio-economic landscape. It is desirable that these upcoming towns be directed away from reproducing the undesirable conditions already existing and continuously being reproduced in the large centres, and should now feature prominently in research and policy undertakings.

In Chapter 3 Kironde evaluates a number of policy approaches adopted by the government to try and deal with problems emanating from rapid urbanisation. The policy approaches evaluated are those relating to: employment and income generation, land and shelter, urban food distribution, urban waste management, urban social services, urban water and urban energy supply, urban transportation and urban security. It is argued that in most cases, success has been limited. There are attempts now to incorporate the private sector and even NGOs and CBOs. Nevertheless, the issue of reform and enhanced good governance is shown to be crucial if headway is to be made in coping with rapid urbanisation.

In Chapter 4 Mascarenhas addresses the crucial issue of migration as a key aspect of urbanisation which, however, has not received as much attention as it deserves because policy makers and researchers have looked at only limited facets of migration. The reasons for, and the destination of, migration are poorly understood, and options and possibilities for participatory development have not been exploited. There has been a natural population census in each of the years 1947, 1957, 1967, 1967, 1978 and 1988. Analysis of aspects of migrations has been made extensively for the 1967 and 1978 census figures, but not for the 1988 one. Unfortunately the National Census slated for 1998 has been postponed. This means that our knowledge on the socio-economic factors of migration, on migration and urbanisation (and especially the differences between large and small towns), on understanding the evolution and development of urban systems, and on understanding poverty in urban areas, and the urban poors' potential for social and economic advancement, will continue to be limited. Mascarenhas proposes a research agenda which will look at spatial aspects, the use of resources and emerging opportunities. Once again the question of good governance is emphasised.

In evaluating the status of governance and urban planning in Tanzania, Lerise (Chapter 5) points out that planning practice is characterised by centralisation of powers and exclusion or limited inclusion of the general population. Planning is still rooted in the concept of land use conforming to a blueprint prepared or approved by experts usually at central government or donor level. The preparation and implementation of these plans lacks transparency and accountability and hardly includes the civil society, the private sector and other stakeholders. In other words, urban planning in Tanzania is characterised by poor governance. He cites cases where local residents' proposals for a land use scheme for their area were ignored by the planners for no sound reasons. In another case, it is shown that action by the civil society in collaboration with officials is helping in directing land use development, and keeping part of a small urban centre clean.

While reforms are going on, these are still mainly institutional. Planners need to be re-educated in order to abandon the top-down and bureaucratic plan production approach and, instead, adopt communicative and collaborative planning where different stakeholders take part in the processes of managing change. Many land use plans prepared by officials remain unimplemented as a result of inappropriate approaches. Urban areas in Tanzania continue to grow haphazardly because urban planning is still rooted in approaches of the past. Research into proper forms of urban planning in a good governance paradigm, as well as successful case studies of the latter are called for.

Chapter 7 by Kessy is somewhat different from those described above in the sense that it is written from an implementation point of view, by somebody who is an activist in community based organisations (CBOs) as pressure groups, as resource mobilisers, and as providers of services and managers of the environment in urban areas. The example of a local CBO that he presents, known as Tabata Development Fund, has attracted local and international interest, and is being emulated elsewhere in urban Tanzania.

Throughout these chapters, concern is expressed at the lack of good governance in the urban Tanzania. The public authorities still work by and large on their own, thus alienating themselves from the people. Resources that people have, including their ability to pay local taxes are not fully exploited and much of the informal activity that takes place is either ignored, or fought by public authorities. In this latter aspect clashes occur frequently between informal sector operators and public authorities. There is a general lack of transparency and accountability on the part of public authorities, who moreover have failed to provide a conducive environment which can enable the various activities to operate for the maximum benefit of society. Powers to manage urban areas are still concentrated with the central government and public authorities. The non-delivery of services, in a situation of lack of transparency discourages the public from paying local taxes.

It is clear that policy makers and researchers cannot say they know it all. A change in attitude on the part of public authorities is required, but then there is need to carry out both theoretical and empirical studies that will inform public authorities on the best approaches to deal with problems of rapid urbanisation. The various chapters have highlighted areas where research is imperative. The last one by Shayo, points out what the other chapters have missed out.

The future of urban management lies in the transformation of management approaches into those that will be locally conceived and implemented, with appropriate support and interaction from public authorities. The interaction and support from the urban communities, local authorities, the government and the private sector will go a long way towards ensuring that sustainable development is achieved.

REFERENCES

Okpala, D.C.I. (1987)
"Received Concepts and Theories in African Urban Studies and Urban Management Strategies: A Critique," *Urban Studies*, Volume 24, p. 137-150.

UNCHS - Habitat (1996)
An Urbanising World: A Global Report on Human Settlement, London, Oxford University Press.

Tanzania (1995)
The New National Land Policy, Dar es Salaam, Government Printer.

2

STATUS OF URBAN RESEARCH AND POLICIES IN TANZANIA

*Suleiman Ngware**

> During the course of history, urbanisation has been associated with economic and social progress... cities and towns have been engines of growth and incubators of civilisation and have facilitated the evolution of knowledge, culture and tradition, as well as of industry and commerce (*The Habitat Agenda*, 1996).

THE URBAN CONTEXT

There is hardly any controversy regarding the important role urbanisation plays in any country's social, economic, political, cultural and technological development. World history has shown that urbanisation is inseparable from a country's social, political and economic progress, the promotion of literacy and education, the improvement of the general state of health, greater access to reliable social services, and cultural and citizen participation (UNCHS, 1996:2). It should be emphasized that democratization is expected to enhance such access and meaningful participation and involvement for civil society actors, for public-private partnership, and for decentralised, participatory planning and management, which are necessary for the urban future (UNCHS, *ibid*).

The most serious and daunting problems confronting smaller towns, municipalities, and cities in Africa in general and Tanzania in particular, include the following: First, inadequate financial resources. Second, lack of reliable employment opportunities. Third, spreading homelessness and rapid expansion of slums and squatter settlements. Fourth, increased abject poverty and a widening gap between rich and poor. Fifth, growing insecurity and rising crime. Sixth, inadequate and deteriorating building

* Associate Professor, Institute of Development Studies, UDSM

stock, services and infrastructure. Seventh, lack of reliable social service delivery e.g. water, health, education, sanitation, transportation etc. Eighth, improper land use and insecure land tenure. Ninth, rising traffic congestion, increasing pollution, lack of green and open spaces. Tenth, uncoordinated urban development and increasing vulnerability to natural and man-made disasters e.g. floods caused by *El Nino* and the recent bombings of American embassies in Dar es Salaam and Nairobi.

The above scenario emphasises a number of important conclusions. First, the problems facing the country's urban people are monumental and threaten the whole social fabric. Second, the enormous problems just mentioned require deliberate and feasible policies that are informed by objective urban research that draws from global experience but is hinged upon Tanzania's social, political, economic, cultural and technological realities. Third, the existing urban problems have seriously challenged the capacity of the state and the urban people themselves to realise socioeconomic development and environmental protection, which are interdependent and mutually reinforcing components of friendly and sustainable development. Lastly, the permanent solution to the country's enormous urban problems require a deliberate shift from urban government to urban governance in which the involvement and participation of civil society occupies and plays an inevitable and strategic role.

In the Least Developing Countries (LDCs) such as Tanzania urbanisation is increasingly and inevitably becoming a powerful social, cultural, economic and political force. It should be emphasized that the Third World in general and Tanzania in particular is urbanising very rapidly. Currently, Tanzania has about 30 million people, 70-75% of whom are working and living in rural areas. Table 2.1 demonstrates this reality.

Table 2.1: *Tanzania's urban population in % and millions*

Year	*1965*	*1980*	*1995*	*2010*	*2024*
Percentage	5.3%	14.8%	24.4%	36.3%	48.2%
Population	0.6	0 2.7	07.5	17.5	35.7

Source: Sida, (1998:11)

The following deductions can be made from Table 2.1. First, even by African standards, Tanzania had a very low level of urbanisation at independence with only 5.3 % of the country's total population living in urban areas. Second, there is rapid urbanisation taking place in Tanzania as evidenced by the increase of the urban population e.g. from less than 1 million people in 1965 to a projected figure of close to 40 million by the year 2024. Third, the percent of the urban population jumped from 5.3 in 1965 to 24.4 in 1995. Fourth, as will be discussed later, whereas rapid urbanisation is an inevitable part of the country's general development process, it over-stretches the human capacity and material resources necessary for the delivery of the needed social services. This leads to increasing urban poverty and decay. In concrete terms this means increasing unemployment, poor and inadequate social, physical and economic infrastructures.

There is no reason to doubt that the growth in the public sector through swift nationalisation (1971–1988) was an important stimulus for urban growth in that period. However, it is the informal sector which led Dar es Salaam to grow more rapidly than the country's regional capitals. There are about 850,000 *machingas* in the city of Dar es Salaam (Keenja, 1998) and these are increasing every day.

At the same time the forced villagisation during the mid 1970s, led to a rapid, rural-urban migration of people wanting to escape from the lack of resources and strong social control in rural areas (Pedersen, 1997: 25). The rapid migration is documented in reference to the frequent, unpopular and notorious attempts by the country's urban authorities to return the inevitable swelling number of street vendors and small informal entrepreneurs to the rural areas.

Indeed, the urban crisis facing countries like Tanzania should be seen as a result of multi-faceted, multidimensional and multi-sectoral factors - both internal and external - that calls for urgent but carefully planned and executed urban policies and programs. The rapid growth of urban centres has led to the inevitable redistribution and sharing of powers not only between the central and local government levels but also, and more importantly, between the different tiers of government and the steadily emerging civil society. This dynamic phenomenon is the essence of the new urban governance. For instance, the new locus of power as a result of the sharing of power and responsibilities named above hinges upon a multitude of dynamic players and actors such as CB0s, NG0s, squatters, various vendor groups, professional organisations, etc. (Ngware, 1997:237–248).

It should be emphasised that the new urban governance paradigm is inherently positive and should be encouraged. It creates the social, political and economic space needed for each sector and actor to play their role while recognizing the absolute need to work together in mutual partnership and collaborative spirit with the state retaining its traditional role of ensuring law and order, thus creating an enabling environment for various societal groups to assume a pivotal, pro-active, and strategic role in improving and sustaining the social and economic wellbeing of the urban people. It is in this sense that the state and society - all stand to gain - by promoting a vibrant civil society. Indeed, we will all lose if civil society is stifled or severely restricted.

To be sure, urban centres in least developing countries like Tanzania are experiencing enormous problems which are negatively affecting their development process, such as the shortcomings of the past development policies, strategies and tactics in which the focus was on the one hand to "promote" development in the rural areas, and on the other to forcibly ensure that rural people are not flocking into urban centres. There is ample evidence to suggest that the current rapid rate of urbanisation which is characterised by acute shortages of adequate shelter as evidenced by dilapidated urban slums, unemployment, and run-down social, physical and economic infrastructure is dialectically linked to the on-going agrarian crisis.

The fast spread of the incurable and fatal AIDS disease and malaria, and the rapid population shift from one base to another, creates not only ecological imbalances and social disturbances but also transforms sociopolitical, economic and cultural relations, markets and internal economies. In a nutshell, Tanzania's urban centres, like the rest in the developing world, demonstrate a very disturbing reality which can be summed up in this equation:

urban poverty and urban decay = urban crisis.

At the same time, the inevitable globalisation process has some repercussions on urban research and policies. For instance, the rapid growth in world trade, foreign direct investment, and cross-border financial flows over the past two and half decades have been the main manifestations of the increasing globalisation of the world economy (Rugumamu, 1998: 2-3). This phenomenon has been magnified primarily by a global wave of economic liberalization e.g. lowering of tariffs or removing them all together in international trade, encouragement of foreign investment, and the non-restrictive deregulation of commodity and financial markets.

It should be emphasised that what falls under the rubric of globalisation is not only contested but is also complex and often confusing. Suffice it to say (for the purpose of this discussion) that globalisation is a historical process which requires us to view it as a structured totality within the various levels of the urban problematique e.g. human practice, politics, information technology, economics, ideology and culture are produced and transformed occasioned by, and resulting in, competing social and political forces asserting power over each other (Rugumamu 1998: 2).

For instance, changes in global industrial production system have altered the international division of labour, dislocating Tanzania's competitive advantage in supplying raw materials and cheap labour. At the same time technological advances are not only changing but are also challenging the position of urban centres in the country. Indeed, cheap labour and large markets are not sufficient factors to maintain the position of Tanzania's urban centres in the global economy and trade. As a result the country's urban centres are in a deep crisis of survival and sustainability. This is a product of the marginalisation of urban centres in the country (Halfani, 1997a: 89).

However, it should be noted that urban growth and development are basically local processes in which local urban actors attempt to develop the local urban area in response to, and in competition with, other regions and large organisations (Pedersen, 1997). It is important to recognise that the re-allocation of national and international resources is vital but it should also be appreciated that no amount of external subsidy is likely to have a lasting impact on local urban development if there are no actors in the communities who are willing and able to exploit the subsidy for the benefit of their urban locality and that of the country (Pedersen *ibid*). The existing knowledge regarding the status of urban research and policies in the country is influenced by a combination of dynamic social, historical, political and economic factors. This is the focus of the next part of the discussion.

HISTORICAL CONTEXT AND PAST EXPERIENCES

Although the discourse on urban governance is a recent phenomenon in the country, there have been a number of policy initiatives, activities and efforts since the country's independence in 1961. The main players and actors include civil society, the marginalized urban poor themselves, central government, local authorities, academics/researchers (both local and international) and some donor agencies e.g. UNDP, World Bank, USAID, European Union, NORAD, DANIDA, Sida etc. According to Halfani (1996:183-200) the early urban studies in the 1950s-1960s were mainly ethnographic surveys, changing configuration of urban places, rural-urban migration etc. It should be emphasised that urban planning and comprehensive urban policy formulation were not treated seriously.

Indeed, urban policy makers and researchers saw urban areas as centres for serving the country's strategy for promoting and sustaining rural development. Hardly did they see them as strategic centres for monumental opportunities for the country's overall economic growth and sustainable social development. According to Pedersen (1997) towns were seen as centres for the distribution of urban services to the rural areas; and being based on innovation, diffusion, and growth-pole theory. Indeed, they were seen as growth points or as the last link in the urban hierarchy through which innovations and modernization trickled down from the large urban centres to the rural areas. Around the 1970s - 1980s, under the impact of the Dependency theory and the Marxist centre-periphery theory, towns increasingly came to be seen as administrative and political centres for resource exploitation, taxation and control rather than as service centres for rural development (Pedersen, 1997: 11-15).

It is in this sense that the 1970s and 1980s (the peak of *Ujamaa* and rural developmentin Tanzania) saw the emergence of government determination to improve its urban knowledge and policy framework through various means including consolidation or opening up of some tertiary institutions such as University of Dar es Salaam, the Ardhi Institute, the Building Research Unit etc. Gradually communication and interaction among academics/researchers was being established.

Halfani's (1996) survey indicates that there were 37 studies on urban issues in 1960s, about 150 studies in 1970s, and there was a decline to 100 studies in 1980s as the social and economic crisis began its long, unprecedented and painful bite. A review of urban research carried out in the country by both state and non-state organisations suggests that concentration was on physical planning, the environment and formal institutional arrangements pertaining to managerial processes (Halfani, *ibid*). The informal sector was, however, not considered as an important strategic ingredient of the urban development process but rather as a mere appendage or an unwanted structural by-product.

Furthermore, findings from research studies between 1960s and 1980s indicate a disturbing phenomenon. First, the state was the major agent determining urban change policy formulation and implementation and evaluation. There is abundant evidence to this effect such as the unilateral abolition of municipal authorities, restricting and stifling of the once vibrant civil society, etc. Second, because the government's focus was on the transformation of the rural sector as its strategy for development, there was little attention given to the development of municipalities and the city, although it is true that Dar es Salaam, Mwanza, Arusha, Tanga, Dodoma, Moshi etc. did receive some attention mainly from multilateral and bilateral donor agencies. Third, it should be remembered that the country's smaller and up-coming towns like Kibiti, Mchinga, Kibaya,

Kibaha, Bariadi, Babati, etc. in Tanzania hardly attracted the attention of researchers, policy makers and donors. The situation is pretty much the same today!

Fourth, the dominance of state-centric approaches to the system of urban governance has also directed research attention to institutional relationships, managerial capacities, effectiveness of the local councils, red tape and bureaucratic domination etc. In a nutshell the top-down bureaucratic manipulation by the state permeated every sector of the social fabric including research and policy orientation. It is therefore not surprising that there were more studies done on rural issues than on urban issues.

Fifth, was the disturbing truth of the fact that urban research agenda and financing of urban research was basically donor driven. The same situation is true today. This is not something to be proud of. The government should consider investment in urban research as an important productive investment for the formulation of effective urban policies. Nevertheless the urban crisis and urban governance issues are about human beings - people - with their dynamic historical, social, political, cultural, technological, psychological, and economic life. Indeed, such attributes have, and will, always remain dynamic and not static. This is true of the whole social fabric - the urban people themselves, the different tiers of the state, academics, researchers, and donors, etc.

At the beginning, research in Tanzania was less analytical and detailed but as time went by, improvements were made. Halfani (1994a:121) has shown that the early urban research was mainly ethnographic in the form of social surveys. But the 1970s saw more rigorous and analytical methods where, not only urban entities of all colour and class were covered but also physical aspects. The former studies were descriptive in nature and were influenced by modernisation theories. The latter studies saw more researchers, more themes and more institutions carrying urban studies. Research orientation was more prescriptive, and more evaluative and analytical and helped to portray a good picture of understanding the process of urbanisation in the country. Studies went as far deep as analysing previous urban policies and programs, urban projects and the technicalities of managing urban areas.

The liberalisation of the late 1980s and 1990s has ushered in the era of state withdrawal from economic and social activities, and the re-emergence of civil society and researchers/academics focusing more concretely on urban *governance* issues rather than mere urban *government*. It is during this decade that local and external urban researchers and policy makers began to take a pivotal, pro-active and strategic role in expanding the urban governance research and policy horizon. A number of important findings could be traced back to this decade. These include, first, the accumulating evidence of the unclear definition of roles between urban officials and the civil society. Second, there was eroded legitimacy because of the long tradition of poor urban governance. Third, urban centres, particularly the upcoming towns, had a weak resource base and a poor revenue collection system. Fourth, findings showed lack of coordination among researchers and policy makers, lack of effective mechanisms for involving the urban people in issues affecting them, weak and uncoordinated dissemination of urban research findings and above all, the inexcusable failure by the actors and players to address the persistent urban crisis that is agonisingly affecting people in the form of endemic urban poverty and notorious urban decay.

There is an emerging consensus among the urban policy makers and researchers that urgent and sustainable measures have to be taken to address the fundamental issues regarding urban governance, poverty and decay if the country is to achieve sustainable development and avoid the unnecessary cost and agony of political strife and social havoc. For instance the dilapidated slums in Dar es Salaam, Mwanza, Dodoma, Arusha, Tanga, Moshi etc. in Tanzania remain not only an economic eyesore, but a socio-political as well as a psychological one. Abject poverty haunts both the rich and the poor in our big towns and municipalities as evidenced by street beggars, street children, prostitutes, high crime rates, (including armed robbery), endemic unemployment and the general sense of urban insecurity which is manifested in urban houses being over-grilled and appearing like voluntary prisons, the mushrooming of security guard companies e.g. Ultimate Security etc. In addition the rapid decaying of the urban environment, poor transportation systems, poor urban governance, corruption, disregard of public opinion, and the inadequate, costly and bureaucratic provision of social services have all made life, and indeed living in the small towns and cities, a constant nightmare (McCarney, 1996:188).

Two more observations need to be made here. First, a permanent solution to the urban crisis just highlighted can no longer be postponed without inducing a social and political catastrophe. Second, while it is true that the magnitude and intensity of the urban crisis are more prevalent in Dar es Salaam, Mwanza, Tanga, Arusha, Mbeya, Dodoma, Moshi etc, it is equally true that the smaller and upcoming towns are following suit although in a smaller scale and intensity. Nobody can deny that smaller towns like Mchinga, Babati, Kibaha, or Bariadi are comparatively and relatively safer towns than Dar es Salaam although they are also increasingly becoming unsafe. Should these upcoming smaller towns inherit the same agonies of the country's big towns? What can the big towns learn from these upcoming towns? What type of symbiotic relationship exists between them? There is no doubt that smaller towns are performing specific tasks within the larger societal system of production and distribution, and are being shaped by the specific social, political, cultural and economic environment of which they are a part (Pedersen, 1997: xi).

Indeed, in a spatial world, smaller towns and their intermediary functions become important links between rural areas and the larger urban centres (Pedersen *ibid*: 1). Small towns are centres for the production and distribution of goods and services to the society as well as for the exploitation and extraction of resources. In addition, they are centres of growth and consolidation of rural non-agricultural activities and for the location of decentralised branches of government administration and services and of small to medium private enterprises.

Small towns are also infrastructural and communication centres within and outside the rural areas and for the retention of some rural migrants who would otherwise flock into the large urban centres, where social costs are often considered to be exorbitant. Indeed, the ability of the smaller towns to provide improved living conditions depends on their ability to transform and expand their economies, and that in turn, depends on the structure of their dynamic links with both the rural areas and the larger urban centres in the country. However, such universal generalizations and prescriptions concerning small town development are simply not valid. It is a known fact that the growth of smaller towns, like that of big towns, does not follow a liner pattern.

FUTURE DIRECTIONS IN URBAN RESEARCH

The future of urban research in Tanzania is promising. As noted earlier, the opening of institutions of higher learning should lead to improved research activities in urban areas. For example, Ardhi Institute has developed further and is now a constituent college of the University of Dar es Salaam. The long-term plan is to transform it into an independent University. Its teaching and research capacity is being consolidated further which will provide an opportunity for training more researchers in urban issues.

The growing number of scholars and the capacities of institutions to conduct research in urban issues is also increasing the diversity of issues under investigation. Apart from the traditional areas of research such as housing and shelter, physical planning and land survey,etc. new areas have been added in the list. Urban agriculture, governance, civil society, gender, crime, urban health, material consumption, social exclusion, house-girls, child labour, commercial sex workers, sustainable urbanisation and such other areas are now being examined by many scholars. With the growing proportion of urban population, the complexity of areas for study is also growing. The country is currently reforming and policies of globalisation, free market, cultural transformation, rapidly modernising technology and the democratisation process will also have an impact on the nature of future urban studies in Tanzania.

It has been shown earlier on in this discussion that, unlike in the 1960s, 1970s and 1980s, urban research and governance in the 1990s has shifted from being state-centric to a multi-faceted, multi-sectoral and multi dimensional paradigm in which various players, stakeholders and actors are the new strategic partners with an enormous impact on how urban governance and research is managed e.g. in terms of policies and research focus and priorities with an ultimate aim of fighting urban poverty and decay. Whereas it is true that the growth of towns is not only inevitable but has also been associated with the urban crisis of an enormous proportion, it is equally true that the urbanisation process offers enormous development opportunities. These need to be thoroughly identified and prioritised. This is the focus of the next section.

FUTURE POLICY AND RESEARCH PRIORITIES

Research in urban issues in the country has been conducted in an unco-ordinated fashion. This has several implications. Urban centres in this country have not been researched equally and even in the researched centres, not all parts of the towns have been covered equally. Moreover, not all aspects of urbanisation/urban areas have been covered equally. In some aspects like urban environment and housing, many researches have been done. Again, researchers in most cases have tended to carry their studies according to the needs of the donors and/or the government. This has tended to pull research towards donor interests rather than situational demands. The above behaviours are found because researches were conceived in a sectoral manner. In the same line, the country lacks a one-stop centre where, all urban researches are coordinated or deposited. Each of the stakeholders has its own priority and objectives and its own research team and funds.

Available evidence suggests that the country's urban centres are confronting complex social, political, cultural, economic and environmental problems. It should be emphasised that the urban people themselves, the government and the various local and external organisations are joining hands in an attempt to tackle the problems. In what follows we provide an overview of future policy priorities and research issues.

First, in 1995 the Tanzanian government adopted a land policy following a report from the Presidential Commission and in 1996 a Draft Land Bill was produced. The government has, since then received critical comments from various civic organizations e.g. TAMWA, TGNP, various professional organizations etc. Although the Draft Land Bill proposes a customary right of occupancy with full legal status, it does not include community institutions like local governments in urban land management. Most powers are with the Commissioner of Lands and the President. There is a need to study the legal, social, and political impact of such propositions. Second, the Draft Urban Policy came out in 1997. It is a comprehensive document that touches all aspects of urban life. However, some areas need a further clarification. For instance the Draft Urban Policy neither defines the role of towns and cities in social and economic development nor articulates the means and ways of attaining the desired objectives. It is not an overstatement to say that the Draft Urban Policy focuses more on the issues of form of urban development than on the content issues of fundamental substance regarding urban development (Halfani, 1997b:19). There is thus the need to clarify how the social, political, economic, cultural and environmental management of the urban centres can be realised and sustained.

Third, the on-going political and economic liberalisation has forced the state to reduce its monopoly of social, political and economic management of the economy and society (World Bank, 1997). Indeed, there is ample social, political and economic space for other actors and players e.g. various civic groups and organisations. This is an area which needs a broad and flexible policy and research framework in order to create a conducive legal, social and political environment so as to remove potential areas of conflict between the state, local authorities and civil society.

Fourth, the Local Government Reform Agenda:1996/2000, provides a firm social and political basis for good governance of the country's local government system including urban governance (URT, 1996). The Regional Administration Act No. 1 of 1997 provides the needed legal framework for devolution of power to the urban and rural councils in the country (URT,1997). Indeed this form of decentralisation is politically desirable in the present social, political and economic environment. However, there is need for the government to speed up the local government reform process and to provide a policy framework that stipulates the governance process at the sub-municipal/city levels where the majority of urban people live and work.

Fifth, the division of roles and relationship between the central government and local government authorities on the one hand and the local government authorities and civic groups and private sector on the other hand, needs clarification. It is true that the Tanzanian state has demonstrated the necessary political will and commitment that allows reasonable political space and emphasises positive collaboration and partnership with the private sector and civil society. Nevertheless there is need for both the legal framework and a new culture that ensures that all actors translate their political commitment to achieving tangible results on the ground.

Sixth, it is encouraging to note that there seems to be good interaction between urban policy makers and urban researchers/academics in the country. This symbiotic relationship should be consolidated. It is emphasized that there is need for researchers/academics to research on burning issues that are urgent and relevant for the urban people and the government. They should also learn how to dialogue and communicate with the policy makers so as to ensure that their research findings and recommendations are accepted and utilized by the urban policy makers.

Conversely, policy makers should also demonstrate courage and willingness to work closely with the local urban researchers and academics. It is amazing to note that donors rather than our own government utilise indigenous urban researchers more! Indeed it is an open secret that some of the recommendations given to the government by donors originate from the same indigenous researchers whose access to policy makers is blocked by bureaucratic red tape and a hostile attitude towards the indigenous urban researchers/academics.

Seventh, that rapid urbanisation is not only inevitable but provides enormous opportunities for the country's sustainable development is no longer a controversial issue. Nevertheless, the manner in which the urbanisation process is taking place in specific urban centres raises some serious research and policy concerns. For instance, do we really know the environmental, social, economic and cultural impact of the rapid urbanisation in up-coming towns like Mchinga, Same, Mombo, Kibiti, Bariadi, Babati, Kibaha et.c? Or do we simply generalise from the experiences of Dar es Salaam, Tanga, Arusha, Mwanza, and Dodoma?

Eighth, a number of institutions and organisations are conducting research on various aspects of urban governance but it is extremely difficult to know who is doing what and indeed what urban research findings and recommendations are available and the possibility of retrieving such useful information. Is it not possible to establish within the Ministry of Lands and Human Settlements Development an urban resource centre where policy makers and researchers can access the information already available?

Ninth, most of the research and policy efforts seem to concentrate on the major urban centres where the country's elites work and live. However, given the fact that rapid urbanisation is the phenomenon now, it is socially, politically and economically desirable that equal attention be given to the smaller and up-coming urban centres where there is a tendency to perpetuate the same urban crisis and management that exists in the bigger urban centres.

CONCLUSION

This chapter has attempted to provide an overview of the status of urban research and urban policies in Tanzania. In the process of doing so, it was found necessary to describe, albeit briefly, the context in which urban research and urban policies have been premised over the years i.e. a reflection of the urban sector and the subsequent urban crisis that is manifested in the form of urban poverty and decay. Second, a quick survey of the historical context and past experiences was made showing the policy trends and shifts in urban research and policies since the country's independence, in 1961. It was shown that urban research and policies were determined by the country's social, political and economic context that was in turn influenced by both internal and external dynamics. Third, a brief discussion was provided on the future urban research and urban policy priorities in the country. It was emphasised that the fundamental issue of urban poverty and decay should be given the most prominent attention and resources by all actors including the active involvement and participation of the urban poor themselves.

In conclusion, the following should be emphasised: First, the policy needs, efforts and resources to identify the poor in the urban setting is of strategic importance since this will create the necessary conditions for enabling the targeting of interventions to benefit the urban poor.

Second, urgent research efforts need to be made to gain objective insight into what it means to be poor in a specific urban context e.g. Manzese, Tandale, Kibiti, Babati etc. and to explore the types and causes of poverty and that findings and conclusions from such studies will form the strategic point of departure for the suitable design and implementation of the necessary and positive interventions.

Third, a constant dialogue between urban researchers/policy makers and the urban poor will provide an important input not only to the comprehensive understanding of the urban problematique but also to the design, planning and implementation of the necessary interventions.

Fourth, comprehensive documentation is needed for proper recording of all urban research findings and recommendations so that urban policy makers can easily access them whenever necessary.

The challenges facing the urban people are enormous, so are the real and potential opportunities. What is urgently needed is the recognition of this obvious fact, political will and collaborative efforts by all policy and research actors, players and indeed the active involvement and participation of the civil society in the urban sector.

REFERENCES

Halfani, M. (1997)
"The Governance of Urban Development in East Africa: An Examination of the Institutional Landscape and the Poverty Challenge," in, Swirling ,M. (ed), *Governing Africa's Cities* Witwatersrand University Press.

Halfani, M. (1997)
"The Challenge of Urban Governance in Africa: Institutional Change and Knowledge Gaps," in, Swilling, M. (ed) *Governing Africa's Cities* Witwatersrand University Press.

Halfani, M. (1994a)
"Toward a New Paradigm of Self Governance" in Rothchild, D (ed), *Strengthening African Local Initiative: Local Self- Governance, Decentralization and Accountability* Institute of African Affairs, Hamburg.

Halfani, M. (1994b)
"Urban Research in Eastern Africa: Tanzania, Kenya and Zambia. Towards an Agenda for the 1990s in Stren, R (ed) (1994) *Urban Research in the Developing World: Africa,* Volume 2 Centre for Urban and Community Studies, University of Toronto.

Halfani, M. (1996)
"The Challenge of Urban Governance in East Africa: Responding to an Unrelenting Crisis," in, McCarney, P.L. (ed), *Cities and Governance: New Directions in Latin America, Asia and Africa,* Centre for Urban and Community Studies, University of Toronto.

Keenja, C. (1998)
"Briefing to Members of Parliament," *Majira,* Friday August 14 Dar es Salaam, p. 1-2

Lerise, F. S. (1991)
"Development of Urban Functions in Villages of Tanzania," *African Urban Quarterly.* Volume 6, Numbers 3 and 4, August and November, (special issue)

McCarney, P.L. (1996)
The Changing Nature of Local Government in Developing Countries. Toronto, Centre for Urban and Community Studies, University of Toronto.

Ngware, S. (1997)
"Civil Society and Forms of Political Participation in Tanzania," in, Hofmeister, Wand Scholz, I. (eds), *Traditional and Contemporary Forms of Local Participation and Self Governance in Africa*, KAS, Johannesburg.

Pedersen, P.O. (1997)
Small African Towns - Between Rural Networks and Urban Hierarchies. Averby, Aldershot.

Rugumamu, S. (1998)
"Globalisation and Liberalisation: A Case Against Africa's Marginalisation," Paper presented at the 1998 Annual Meeting of the African Studies Association Chicago: October 30 - November 1, Institute of African Affairs, Hamburg.

Sida (1998)
"Knowledge Base on Urbanisation in Tanzania: IDS Draft Proposal," Dar es Salaam.

URT (1996)
Local Government Reform Agenda, 1996-2000. Dar es Salaam, Government Printers.

URT (1997)
"The Regional Administration Act. Number 1 of 1997." Dar es Salaam Government Printers.

World Bank (1997)
World Development Report: The State in a Changing World. OUP, London, & New York.

UNCHS-Habitat 11 (1996)
The Habitat Agenda: Goals and Principles, Commitments and Global Plan of Action. UNCHS-Habitat, Nairobi.

3

RAPID URBANISATION IN TANZANIA: THE GOVERNMENT'S COPING STRATEGIES

*J.M. Lusugga Kironde**

INTRODUCTION

Over the past 40 or so years, urbanisation - that is the proportional change between the population living in rural areas and that living in urban areas - has become one of the most important trends in human settlements development. This is particularly so in the developing countries of Africa where the rates of urbanisation are the highest in human history despite the fact that Africa is the least urbanised of the continents. Initial theorisations of this phenomenon hinged on the concepts of adverse "push factors" in the rural areas, and "attractive" pull factors in urban areas. Earlier solutions were therefore seen as taking development to the rural areas, and preventing migration to, and controlling residence in, urban areas as a way of stemming rapid urbanisation. Attitudes have changed over the past 20 years and urbanisation has been accepted as an inevitable qualitative change in the distribution of national populations. In any case, rapid urbanisation cannot be wished away.

Even though urban areas are no longer offering "ready-made" employment in the public sector or in the industrial sector, they are seen as possible engines of development, creating diverse opportunities for earning income, ensuring production and distribution at a low cost, and generating all kinds of markets for both urban and rural activities. They also have a tremendous potential to combine safe and healthy living conditions and culturally rich and diverse lifestyles with remarkably low levels of energy consumption, resource-use and wastes (UNCHS, 1996).

* Associate Professor, University College of Lands and Architectural Studies (UCLAS), UDSM.

Rapid urbanisation in developing countries has presented governments, both national and local, with a demanding situation. Many governments approached the problems of urbanisation using concepts received from the colonial era and applicable to developed countries. These included comprehensive government regulation of urban development and public provision, usually at subsidised prices of most urban infrastructure and services. Political considerations too have been important in determining the location of power, whereby the tendency has been the concentration of authority in the central government. Scarcity of resources, both at local and national levels, coupled with macro-economic problems that many governments have been facing in the past decades has limited the leverage that the public authorities have to cope with rapid urban development. Issues of governance, such as the relationship between the central and local governments, between governments and civil society, and within civil society itself, have had important connotations for approaches to dealing with rapid urbanisation and development. In most instances, governments have failed to cope with the demands growing out of this rapid urbanisation and are being forced to rethink many of their earlier policies. There is national and international concern that the urban question must be addressed in a way that will ensure access to resources by all citizens, liveable and sustainable environment, rising standards of living, peaceful coexistence among diverse groups, cultural enrichment, and democracy.

PERIODISATION

African urban areas began to grow rapidly after World War Two, and more so after independence, around 1960. Before then, urbanisation was slow and highly regimented by the colonial government, and urban areas were considered to be only for non-natives and those formally employed. Many observers and commentators on urbanisation in Africa recognise three phases since the independence period. The 1960-70 period is termed the modernisation phase, when African governments in the wake of independence were concerned. with plans to develop their economies including the attempts to uphold the "modern" urban standards left behind by the colonial regimes. "Back to the land" policies were undertaken in many urban areas to try and control urban growth (Mabogunje, 1990).

The 1970-80 decade is considered to be a phase when failure to directly manage urban development was accepted and recourse made to "aided self-help" policies, exemplified by undertakings such as sites and service schemes and the upgrading of unplanned areas (Stren, 1992).

The third phase runs from 1980 to the present, and is characterised by many African cities getting confronted by both an economic and a managerial crisis. Adverse economic problems affecting most African economies have forced new conceptions of development challenges. The period has brought into focus the stark reality that African cities cannot hope to develop on the Western industrialised city model. Instead, many African governments are being advised to adopt or are trying out "enabling approaches," and there is the advocacy to redefine the African city to encompass self-help activities, urban agriculture and informal sector activities (Pugh, 1992). Looking back to the 1960s, African cities of the 1990s have changed at least in four major aspects namely: their size, the quality and distribution of public services and infrastructure, their employment base and their spatial organisation or morphology (UNCHS, 1996).

In terms of size, the population of many urban areas in Africa has grown tremendously in the past three decades. Growth has been most spectacular in the largest (or primate) cities but growth rates in these large cities have somewhat declined since the 1980s. Medium sized towns are now growing very rapidly, perhaps reflecting the problems that face urbanites in the largest cities and national economic problems in general. While during the 1960s rural to urban migration was the most important component of urban population growth, natural increase is now a major element.

Another trend of the 1990s is the deterioration of services and infrastructure. This is an inevitable result of economic stagnation in the wake of rapid urban growth. Under such a situation, the resources necessary for roads, sewers, water systems, schools, hospitals and plant and machinery for waste management have failed to keep up with the needs of the population (World Bank, 1994). This trend also emanates from problems rooted in inefficient management and poor governance. While there is some clear differentiation between the living conditions of the small groups of upper-level managers, foreign diplomats, senior politicians and successful businessmen on the one hand, and the growing number of poor urban dwellers on the other, there is also a general deterioration in the public services and infrastructure. Thus for example, in many cities, water and electricity have to be rationed, roads are full of potholes, and garbage is left to rot on the street.

Changes in the labour market and income generating opportunities have been dramatic since the 1960s. Then, employment opportunities for both skilled and unskilled labour were abundant in government, the parastatal sector, in import substitution industries, and in a buoyant private sector. The expansion in education as well as population growth has created a large supply of labour which must make a living, in an epoch when parastatals are being privatised and the public sector is contracting and its staff being rationalised. The nascent local private sector is also facing fierce competition from outside as a result of economic liberalisation. Even university-trained professionals are finding it difficult to get employment. This has resulted into the growth of what is referred to as the "informal sector" an omnibus term covering all aspects of unregulated or semi-regulated economic activities ranging from street traders, hawkers and food vendors, to small scale businesses, workshops and privately-owned public transport vehicles (Becker, *et al*, 1994). The African city has moved from the overly "regulated city" of the colonial era, to the "self-help city" of the 1970s to the "informal city" of the 1990s.

The city form has also changed. The central business district (CBD), clean with wide streets, was the focus of urban life in the 1960s. Today the CBD is poorly maintained and getting more populated with small-scale street vendors and hawkers than in the past. More and more of the population is moving to the periphery where land is much easier to get and where unregulated activities can be carried out. Planned development, a tenet of the 1960s, is no longer the norm. Spontaneous urban development is creating unplanned settlements everywhere. Not only is urban sprawl stretching the capacity of governments to provide infrastructural services such as piped water, electricity, sewerage and roads, and increasing the cost of services such as education, health and transport, it is also a result of this failure, thus creating a vicious circle.

All the above suggests the need to develop new approaches to the planning and management of African cities, if they are going to survive as viable social and productive entities in the 21st century. Among the crucial demands imposed by rapid urbanisation to which governments are forced to respond are those related to: employment and income generation opportunities, land, housing, food distribution, waste management, social infrastructure such as schools and health facilities, water and energy supply, transportation facilities, and urban security. At the same time rapid urbanisation requires that appropriate forms of governance be evolved that will take into account the diverse nature of urban residents, and the need to meet their specific demands. This chapter broaches policy approaches taken in Tanzania to address these issues.

EMPLOYMENT AND INCOME GENERATION OPPORTUNITIES

It is now generally accepted that the primary motivation for people moving to, or remaining in, urban areas is to seek employment or other income generating activities. The Tanzanian government's policy towards employment was, in the 1960s and 1970s, guided by the belief that there was sufficient income earning opportunities in the rural areas. In 1976, the government launched *Operation Kila Mtu Afanye kazi."* (Every able-bodied person has to work). Those identified as being unemployed were required to either go back to their rural areas, or to go into rural settlements identified around major urban areas like Dar es Salaam (Magoho, 1991). The level of force used notwithstanding, this repatriation policy did not succeed. In 1983 the *Human Resources Deployment Act* was enacted and empowered the government to look into, and find a solution to, the question of unemployment in urban centres. The policy became known subsequently as the *Nguvu Kazi* policy. Considerable responsibility was put on local authorities which were required to form Human Resources Development Committees to plan for employment generating projects and give guidance regarding the undertaking and execution of self-help schemes. The Minister for Labour was empowered to make proper arrangements for the repatriation and subsequent employment of the unemployed urban residents.

Although the *Nguvu Kazi* policy had tendencies of repatriating people from urban areas, and also regarded agriculture as the solution for urban unemployment, it was nevertheless a progressive policy in the sense that many small-scale operators were issued with business licenses on easy terms. Similarly, bye laws such as the Dar es Salaam City Council (Hawking and Street Trading) bye-laws of 1987 as amended in 1989 and 1991 were passed and legalised many informal sector activities provided a license was obtained and trading was carried out on approved premises.

The success of the *Nguvu Kazi* policy is limited because of a number of reasons. One, the role of the various authorities: the Ministry of Labour, regional authorities and local authorities was not made clear and many had overlapping and conflicting powers and responsibilities. Two, the envisaged outlets for the poor in the form of agricultural activities on land allocated at the periphery of major urban areas collapsed. Many of these soon filtered back into urban areas. Three, it became apparent that many *Nguvu Kazi* licenses were awarded to people outside the targeted group, many being high-income households. Four, the number of informal sector operators qualifying for licensing became too high with more and more migration to urban areas. Thus many activities grew up or were conducted without any licensing (Wenga, *et al,* 1995). For example, Dar es Salaam stopped giving *Nguvu Kazi* licenses in 1993.

Current observations in many urban areas in Tanzania suggest that the *Nguvu Kazi* initiative is no longer of significance in determining policy with regard to employment. A major headache that many authorities are facing is the proliferation of many informal sector activities most of which are not controlled. The National Informal Sector Survey of 1991 calculated that 56% of the total urban employment was in the informal sector (Table 3.1). The majority of informal sector operators (74%) are self-employed while 26% are employers (Tanzania, 1991).

Table 3.1: *Total informal sector employment by sex and geographic area*

Sex	*Dar es Salaam*		*Other Urban areas*		*Rural areas*		*Total*	
	Number	*%*	*Number*	*%*	*Number*	*%*	*Number*	*%*
Male	190971	60.4	354,526	55.9	985597	69.4	1531094	64.6
Female	124587	39.6	279,619	44.1	433680	31.6	828286	35.4
Total	315958	100.0	634,145	100.0	1419277	100.0	2369380	100.0

Source: Tanzania (1991)

According to the Ministry of Labour and Youth, 700,000 people enter the labour market annually, but only 30,000 jobs are available in the formal sector. In Dar es Salaam, 1.2 million people are estimated to be able to work, but only 0.6 million are employed in the formal sector, and the figure is falling as a result of massive redundancies in government and the parastatal sectors, and, as many industries close down or work under-capacity. It is estimated that 300,000 people enter Dar es Salaam annually. Most of these end up earning income in the informal sector.

A notable activity in most urban areas has been street trading which has grown tremendously in many urban areas over the past decade. The authorities' response to street trading has been of utter hostility and there are many occasions when the authorities and street traders have clashed, with the authorities wanting to remove traders from certain sections of the urban areas. In almost all the cases, when the authorities clash with street traders, the former destroy or confiscate the traders' wares and operating tools; sometimes, the traders are themselves apprehended, and fined or imprisoned. The authorities' hostility towards the traders is based on a number of premises:

- That street traders are a public health hazard and a nuisance who contribute considerably to waste generation and littering in the urban areas and who inconvenience pedestrians as well as business premise owners in areas where they are operating;
- That street traders are law breakers who operate without licenses, pay no taxes and are suspected thieves and swindlers; and
- That street trading is not the ideal option for utilising the energy that

many of these traders have; which energy could be more usefully deployed in agriculture, for example.

Many times, street traders find themselves intertwined in the political jostles of the country where some politicians would come out supporting them and others fighting them. For example, in April 1996, the Minister for Labour and Youth stated that government policy was to help people to employ themselves; and was against the repatriation of people to rural areas, since urbanisation was the propeller of development. He pointed out that plans were under way to integrate street traders and other micro-businesses into the economies of urban areas and thus discontinue their habitual harassment. Less than a week later, the Prime Minister ordered authorities to remove all unemployed youths and street traders from urban areas and send them back to the rural areas and "get them to work to contribute to development" (Kironde, 1996).

Governments have not found a solution to the rapidly growing armies of youths who have no formal employment and who have to make a living by undertaking all sorts of informal activities. Steps to provide them with specific operating premises or locations have always proved a short-term palliative; soon informal sector operators go back to those areas where they are not wanted, or invade other areas.

The Dar es Salaam City Commission (DCC) has been accused of mercilessly demolishing kiosks and business premises of petty traders on a number of pretexts including beautifying the city. Many of these petty traders - some of whom have complained of having loans to service - lose their wares, capital and livelihood. In a few cases the city allocates alternative land but this is usually considered unsuitable, lacking markets or customers for example. A more positive role has been reported from the municipality of Moshi (Box 1).

The Dar es Salaam City Commission has tried out a number of what could be regarded as positive initiatives. With the assistance of the National Income Generation Programme (NIGP) two modern markets have been constructed in Dar es Salaam; one at Makumbusho and the other one at Temeke Stereo aimed at accommodating street and other petty traders. Between them these markets are to provide about 400 trade kiosks and 600 stalls. Nevertheless according to the DCC, there are some 850,000 petty traders in the city. Thus the solution provided by the two markets is a drop in the ocean. Moreover, problems have already manifested themselves in the running of the Makumbusho market, which is operational. The rates levied by the DCC for stalls and kiosks are considered to be too high by the traders. As of 20th August 1998, 238 out of 298 stalls at Makumbusho market had no traders despite the DCC reducing the rates by more that a quarter. Existing traders complain of a number of things including lack of customers.

BOX 3.1: MOSHI ADOPTS A PROGRESSIVE POLICY TOWARDS THE INFORMAL SECTOR

In a situation of confusion on what should be done about petty traders in urban areas, Moshi Municipal Council decided to approach the problem positively. The Council does not see these traders as a "bunch of hooligans roaming the streets, blocking traffic with their goods spread on the sidewalks." On the contrary, they are regarded as people creating self-employment who also contribute to the GDP. Rather than chase them around, the municipality decided to hold a dialogue with them and to settle them on approved sites, to give them licenses to enable them to carry out their businesses lawfully and without harassment and to look for funds to give them loans. The Municipality is working closely with the Moshi Centre for Informal Sector Promotion (CISP) which raises funds to train youths in various skills and to provide them with loans. Other authorities should perhaps copy the Moshi approach although with a population of only 150,000 people and about 300 street vendors the scale of the problem in Moshi is different from what it is in a large city like Dar es Salaam.

The NIGP has supported the construction of a similar ultra modern market in Arusha, where the Kilombero market is earmarked to take some 2000 traders. All the same, it would appear that the construction of ultra modern expensive markets is probably not the solution to the problems of petty traders in urban areas.

Apart from the construction of markets, other initiatives undertaken by the DCC and other urban areas, usually in collaboration with NGOs have included having open-air markets, closing some streets to traffic, and giving loans to, and training petty traders and other small scale operators. The impact of these on the magnitude of the problem has been limited (Swai, 1998).

Overall, however, urban authorities in the country have not been very active at promoting economic development in their own localities. The problem is that of a national nature. Many employment opportunities are dependent on national policies on investment, taxation, privatisation and so on. In 1996, the government announced the launching of a National Youth Policy, which among other things, aimed at creating employment for them. This, however, is yet to be put in practice. In 1997, a National Employment Policy was launched, but its results are yet to be seen.

It is nevertheless realised that the problem of employment and income generation in urban areas is complex, and solutions must be considered at national and local levels. Some of these include the need to diversify the employment and income generation base including reviving industries, encouraging entrepreneurship and making agricultural activities lucrative. There is need to encourage the growth of small urban areas and trade centres, which would act as markets for agricultural produce. On their part, urban authorities need to encourage business undertakings in their areas by creating a conducive environment for all investors, large and small, by undertaking deliberate business promotions, investing in human capital, providing land and other infrastructure and streamlining and easing the regulatory framework (e.g. easing the licensing process). By improving their revenue base, urban authorities will be in a position to invest in social and physical infrastructure, which in turn will have a positive effect on employment and income earning prospects.

LAND FOR URBAN DEVELOPMENT

Together with the proliferation of informal sector activities, the problem of providing shelter and land to the growing urban population was realised in the 1960s. The reaction then was to try and provide complete shelter through the public sector, to fight the growth of unplanned areas, to control house prices through rent control, and to manage land development through the provision of planned, surveyed and sometimes serviced land.

In the light of the world-wide experience of the failure of the above policies, the government, with advice and support from the World Bank, opted for sites and services schemes and the upgrading of unplanned areas in a number of urban areas including Dar es Salaam, Mbeya, Mwanza, Tanga, Morogoro, Iringa, and Tabora. These undertakings remained few and apart and could not cope with the increasing demand for land and shelter.

Provision and Management of Urban Land

Currently, the demand for planned land far outstrips the official supply so much so that, according to the Ministry of Lands and Human Settlements Development (MLHSD), 65% of all the urban population nation wide lives in unplanned areas, where land is obtained from the informal land delivery system. A glimpse at the poor performance of the public sector in providing land can be gauged from the figures given in Table 3.2 which shows that in both 1990 and 1995/95 less than 5% of the demand for surveyed land was met. Moreover, the number of plots surveyed has gone down drastically from 21,000 that were surveyed in 1972 to less than 7,000 that were surveyed in 1995/96.

Table 3.2: *Number of applications for surveyed plots and number of plots surveyed, 1990, 1995/96.*

	Applications for plots surveyed	*Plots Surveyed*	
Year		Numbers	as % of applicants
1972	N.A.	21,000	N.A.
1990	151,600	7,114	4.7
1995/96	157,000	6,700	4.3

Source: Department of Survey and Mapping, MLHSD 1997.

The case of the city of Dar es Salaam is also illustrative. Between 1978/79 and 1991/92, 261,688 applications for plots were received by the DCC, but only 17,751 plots (i.e. 7% of the applications) were surveyed and allocated (Kombe, 1994). It is calculated that between 1972 and 1994, the annual reported demand for planned plots was 20,000, yet the average number of plots allocated by the authorities in Dar es Salaam was only 2,000. In cases where land has been planned and surveyed, this land has lacked the usual infrastructure such as roads, water and electricity networks and drains and sewers.

The government's policy for coping with this demand for urban land has generally been to believe in the possibility of mustering sufficient resources to execute planning schemes large enough to meet the needs of rapid urbanisation. Concurrent with this belief has been a general disregard of the parallel system that is providing land for urban needs, although this has its own limitations. The current public urban land management system has a number of problems, which inevitably limit the output of planned land as indicated in Table 3.2, and which lead to the growth of unplanned settlements. These include the following:

- The general planning process (including land acquisition, planning, surveying, and registration) is too slow compared to the speed of urbanisation. Moreover, most planning powers are concentrated within the MLHSD. This has on a number of occasions resulted into power overlaps and clashes between local authorities and the central government;

- Low productivity of the public sector in terms of planning and surveying land;
- Dearth of land on which to execute planning schemes, since much of the land surrounding urban areas is owned in one way or another under customary or quasi-customary tenure. Acquiring this land is unpopular and politically sensitive particularly since the government only pays paltry compensation (which ignores the value of land) and even this is rarely paid on time. With time, land at the periphery of urban areas has been acquired and developed to high standards by rich or socially high ranking people, making the question of land acquisition more difficult (Mwamfupe 1994);
- Dearth of resources on the part of the government generally, but also as a result of allocating planned land at very much below its market value and below what it costs the government to produce such land. The government's argument for under-pricing land is to protect the interests of the poor, although in actual fact, it is the rich and socially powerful who end up getting planned land usually at the expense of the poor who get displaced, and who acquire land from the informal land delivery system at a price many times higher than that of official land.

In order to muster sufficient resources, the government has taken some steps to increase the price of land (e.g. by increasing land rent). In 1991 the Minister for Finance announced that the government was setting up a revolving fund for planning and servicing land which would be funded by charging the full cost of land surveying, and by auctioning choice plots. The fund is yet to make a debut or to have any appreciable impact on the supply of urban land. It is significant that this fund was to be centralised and not replicated at local government level.

Meanwhile, most seekers of urban land get it from the informal sector by buying it from its bona fide owners. This process has its own problems, namely:

- Such land has no planning certainty and many times conflicts have occured between the land occupiers and the government where the latter could argue that the land had not been developed according to the planned use;
- Land is developed haphazardly some times to very high densities without the standard provisions for public physical and social infrastructure such as land for roads, open spaces, schools and so on. This creates a hazardous environmental situation unsuited to modern human habitation;

- Some of the land occupied particularly by the poor is marginal, located in valleys or on steep gradients, or in polluted locations. In some cases, land used as burial grounds has been invaded for construction, causing cultural and social strains.

The government policy towards the informal land delivery system has generally been to ignore it. However, in some cases, there has been moves and attempts to remove these settlements but these moves have got entangled in political quagmires and have rarely been implemented. This has been attempted in the case of valley dwellers in Dar es Salaam and steep slope dwellers in Mwanza. The case of Dar es Salaam has been most interesting. The DCC made a survey and identified over 5,700 houses to be demolished from valleys liable to flooding (Table 3. 3).

Table 3.3: *Number of valley dwellers' houses earmarked for demolition by the DCC*

KINONDONI ZONE		ILALA ZONE		TEMEKE ZONE	
Ward	No of Houses in valleys to be demolished	Ward	No of Houses in valleys to be demolished	Ward	No of Houses in valleys to be demolished
Magomeni	373	Upanga	78	Mtoni	423
Ndugumbi	157	Jangwani	283	Miburani	401
Hana Nassif	68	Mchikichini	163	Keko	120
M'nyamala	385	Tabata	853	Kurasini	80
Kawe	26	Buguruni	338	Mbagala	240
Mzimuni	183	Vingunguti	354	Yombo	
Kigogo	540	Ukonga	18	Vituka	91
Ubungo	97	Ilala	35		
Makurumla	170				
Mabibo	64				
Kundudu	120				
Kibamba	98				
Zonal Totals	2,246		2,122		1,356
Grand Total					5,723

Source: DCC, July 1998.

This created a political saga. Many valley dwellers said that flooding was not their major problem compared to lack of land, preferred locations, and lack of resources to build again. Many claimed to be lawful owners of the land, e.g. from customary or even statutory laws. Many blamed the government for having left them to spend their resources on putting up buildings, only to threaten them later with demolition. A problem also cropped up as to where to relocate them since this involved the displacement of other people. It would appear that overall the government is finding it difficult to remove people once they have settled in an area.

Since 1989, the MLHSD has prepared layout schemes for many unplanned areas, which could form the basis of regularising land in these areas. These schemes have been difficult to implement as they differ considerably from what is actually on the ground, and since they are prepared with no consultation with the land occupiers (Kironde, 1997a). The government has not only failed to ensure an adequate supply of planned land for residential use, but has also failed to meet demand for other uses such as industry, public use, waste disposal and so on. A notable failure has been with respect to burial grounds in major urban areas. In Dar es Salaam, many existing burial grounds are full and have been declared closed by the authorities but continue to be used.

Some Effects of the New Land Policy

Problems of getting land for urban development were instrumental in prompting the government to think out a New Land Policy (Tanzania, 1995). The New Land Policy recognises that land has value. A new "more explicit and full bodied 'Land Act' which will be more readily understood and responsive to the needs of modern Tanzania in the context of the liberalised economy and emerging land market" was passed by Parliament in 1998.

There are many positive aspects in the new Land Act, 1999. These include the recognition that land has value, the setting out of provisions aimed at dealing with land under customary tenure, providing powers for validation of certain dispositions of land carried out without approval and providing powers for dealing with irregular settlements in urban or peri-urban areas. However, there are major shortcomings in the Land Act, which can make the management of urban land difficult (Kironde, 1998).

First is the concentration of powers of urban land administration in the Commissioner for Lands at the expense of local governments. Local governments are prevented from giving offers for land, and the Commissioner has powers to direct land officers with local governments to act according to his/her orders contrary to the orders of these local governments. All applications for granted rights of occupancy can only be made to the Commissioner, and any change of use over, or disposition of land carrying a granted right of occupancy must be sanctioned by the

Commissioner for Lands. The Commissioner for Lands can consult local authorities in whose area of jurisdiction the land in question is, but he/she need not take their advice.

The logic behind such concentration is possibly to reduce conflicts over powers of land administration. Nevertheless, the ideal situation in urban land administration is where powers are devolved to as a low level of local government as possible. Certainly local government need to be given an active role in land administration, particularly since, parallel with the New Land Policy, there is a Local Government Reform Agenda (1996-2000) going on and aimed at creating strong, efficient and democratic local governments in the country. These cannot be created if they are shorn of most powers of administering and managing their lands.

Secondly, and likely to create a very serious situation for urban development, are the provisions to do with irregular settlements in urban or peri-urban areas. It is the Minister for Lands who, of his/her own volition or if requested, has the powers to consider that an area be declared an area of regularisation. It is the Minister (or the Commissioner for Lands) who then moves into this declared area and interacts with the stakeholders, prepares a regularisation scheme, and implements it. The scheme can only be approved by the Minister. The local authorities and the people involved are onlookers or stakeholders to be consulted. In all honesty, powers to regularise unplanned areas should be concentrated in local authorities as well as in neighbourhood organisations.

Thirdly, the new Land Act completely ignores the various actors that make the informal land delivery system to work. The land owners, the land seekers, the middlemen, the land developers, the local formal and informal institutions that oversee and guarantee land sales and land tenure, and regulate or fail to regulate land development and that resolve disputes, as well as the emerging community-based organisations are all not given any substantive role, although many studies have shown the key roles that these play in administering urban land (Kombe, 1997).

Fourthly, although most landowners get it through buying, and although the New Land policy recognises that market transactions in land are common, the new Land Act falls short of openly recognising market transactions, except in the sense of validation. This is static and reactive, in a situation where dynamism and pro-action is required. There are no specific steps included to encourage the open selling and buying of land, and there is no institutional support for this process especially in the informal land delivery system.

Fifthly, the question of customary tenure is left vague, as has always been the case in the past. Many people who acquire land from the informal land delivery system especially in the urban fringes assume that they hold land under customary tenure. In some cases such transactions have the blessing of state institutions such as political parties or village governments. The colonial conceptualisation of customary land tenure excluded individual land ownership, and the selling and buying of land. Yet, most urban land obtained informally is obtained through purchase, and is owned by individual households. In a recent case in Dar es Salaam involving the expansion of a road, many people lost property worthy of millions of shillings through demolitions. The government, (through the Attorney General) argued that most of the affected could not claim to be holding land under customary tenure since many were immigrants to that area from other parts of the country. This is a statement that potentially turns the majority of urban land owners into squatters and should have been addressed by the new Land Act.

Because of the failure to address many of the problems highlighted above, public authorities in Tanzania are, and will continue, to be more and more marginalised in the whole process of urban land administration and management whereas in fact, a positive role of the public sector is essential for orderly urban land development.

Housing and Shelter

As pointed out above, in the 1960s the government had hopes of solving the urban housing problem through the provision of completed housing through such organisations as the National Housing Corporation. The contribution of the NHC to the total housing stock has remained minuscule and has been falling with time (Table 3.4). From 1987/88 to the present, NHC's contribution to new housing stock has hovered around 30 units per annum.

Table 3.4: *Units completed by the NHC 1962/63-1989/90*

Year	*Units completed*
1962/63-1971/72	11,632
1972/73-1981/82	4,098
1982/83-1989/90	234
Total	15,964

Source: NHC Records (various years)

In 1982, the government published a National Housing Policy hinged on the expectation that employers shall set aside some of their money to construct housing for their employees. Again this had little, if any, impact on the number of new houses. For example, according to the Ministry of Lands, there is a shortage of 1.5 million houses in the country's urban centres, and this is expected to rise to 2.2 million by the year 2000. This is possibly an exaggeration since much of the housing provided, and likely to be provided in the informal housing market, is possibly not fully taken into consideration. Nevertheless, there is concern that many urban residents are living in run-down and overcrowded housing conditions. It is estimated that between 20 and 40% of all urban dwellers live in overcrowded conditions where, moreover, the water and sanitation conditions are very poor.

Current evidence shows that most urban dwellers are tenants, renting property from private landlords in unplanned areas. The only government reaction to this situation is the Rent Restriction Act of 1984, which is not dynamic and which has generally sought to depict landlords as exploiters who must be prevented from milking or molesting their tenants. Most powers over housing administration e.g. determining rent, getting vacant possession and so on are taken from the landlord to the various tribunals created by the Act. A recent study found that the Rent Restriction Act is ineffective since many landlords and tenants ignore it (Kabwogi, 1997).

Another aspect concerning the provision of urban housing has been the question of housing finance. In 1995, the only mortgage bank in the country, the Tanzania Housing Bank, collapsed. All the same, it had financed only a tiny number of houses constructed in the country. The government has plans to re-establish another mortgage bank, although even if this would be in place, it is likely to finance only a limited number of houses in the high income brackets. Many households therefore finance their own housing from personal savings. This means constructing a house over a very long period, sometimes running into decades. The question of buying a house does not, by and large, exist in Tanzania.

A minor aspect of housing finance that is worth mentioning, is the housing finance facility for civil servants. This was established in 1992. It has so far benefited 287 civil servants, mainly enabling them to build their houses but in the case of some (73), enabling them to buy houses from the NHC. A total of T.shs. 454million has so far been loaned out (Tanzania, 1998d).

Overall, the government needs to evolve a comprehensive policy with regard to urban housing to cover both the drive for owner occupation and renting and to cover issues of housing finance. In particular investors, large and small need to be encouraged to put their money in housing for owner occupation and for renting. Policy initiatives are also required to improve the quality of the current housing stock and its environment in all parts of the urban areas, especially the older parts, the slums and unplanned settlements.

FOOD DISTRIBUTION: URBAN MARKETS

The question of urban food supply is complex. Urban growth modifies food purchasing habits, and makes existing market areas and infrastructure inadequate. We deal with just one aspect of urban food supply, that is the question of urban markets.

In most urban areas, local governments sometimes with the support of central governments do put up wholesale and retail markets to ease food distribution. Nevertheless, in many urban areas the authorities have failed to provide sufficient markets to meet demand. Many official markets get overcrowded and spill over to neighbouring land. Also services to these markets such as waste removal, sanitary services do not exist. Many markets are genuine health hazards. At the same time, many spontaneous markets have developed in many urban areas. For example there are 75 official markets in the city of Dar es Salaam. However, there are another 52 markets, which developed spontaneously.

In many urban areas markets are constructed and managed by local authorities who charge traders a fee for using the markets. Recent studies have shown that market fees are low and are poorly collected so that the local authorities do not get sufficient revenue to manage the markets. The rate for Dar es Salaam is Tshs 300/= a month a stall (CTI, 1997). In response to these findings as well as the need to raise more revenue for local governments, many urban authorities have put up market dues. Unused to paying realistic dues, and unconvinced about the services offered by the local authorities, traders have resisted the hiking of rates in such urban areas as Sumbawanga, Mbeya and Dar es Salaam (Box 5.2).

Markets have also an added element in that they provide premises for economic activity for many urbanites and this has connotations for urban poverty alleviation. Inadequate or unacceptable premises for informal ector activities has led to the growth of unregulated street trading in many urban areas. Partly in order to address this question the Dar es Salaam City Commission with assistance from the National Income Generation Programme has put up two new ultra modern markets in the city at Makumbusho and at Temeke Stereo. The former is already in operation but problems have been experienced in a number of areas. The selection of who should be allocated a stall brought about some considerable controversy; and traders have considered the fee set up by the DCC as being exorbitant and totally unrelated to the turnover they were experiencing. Also the market is not attracting sufficient customers because of other informal markets to which customers can go. The DCC has recently been forced to revise the fees downwards.

BOX 5.2: DAR ES SALAAM MARKET TRADERS REFUSE TO PAY HIGHER MARKET DUES

The Dar es Salaam City Commission (DCC), in contradistinction to the suspended Dar es Salaam City Council has been making efforts to update the various charges due to it and to effect collection. Partly in view of improving the finances to be able to maintain markets the DCC decided to hike market stalls and kiosk charges from Tshs 200/= and 500/= a month to Tshs 300/= and Tshs 1,000/= a day per stall and per kiosk respectively. The businessmen, through their organisation, protested and took the issue to court. Up to now no verdict has been reached upon and the DCC is collecting nothing from its markets. In a number of cases, businessmen have themselves put up their own organisations to ensure that markets are serviced. In the case of its newer markets, the DCC decided to leave the management of the markets to private firms. The arrangement is producing mixed results with traders complaining of high charges (Tshs 20,000/- and 6,000/= per month per kiosk and per stall respectively, plus 50/= per head per day for the use of toilet facilities) and poor service delivery.

An interesting initiative adopted recently by Mwanza Municipal Council is to privatise two of its major markets. The management and revenue collection from these markets have improved tremendously. The monthly revenue now collected is T.shs. 1.55m. from the Central Market (up from the previous figure of T.shs. 0.52m. annually). The amount collected from Mwaloni market is T.shs. 18m per month, up from a figure of T.shs. 4.5m. annually (Kironde 1997b).

In terms of policy and research considerations, it is clear that local authorities especially in the bigger urban areas have not invested sufficiently in urban markets. In a number of cases land earmarked in planned areas for markets has subsequently been put to other uses. Besides, in many unplanned areas there is no official land set aside for markets. Urban markets management needs to be addressed especially in terms of bringing in the private sector, collecting realistic revenue and ensuring that services are delivered; and ensuring investment in new markets. Spontaneous markets that have sprang up to fill in the shortage of official markets need to be looked into with the view of bringing them into the orbit of the authorities' management set ups.

URBAN WASTE MANAGEMENT

Solid Waste Management

Unlike the case with rural areas, solid waste management is a necessary undertaking in modern day urban areas. Municipal solid waste include refuse from households, waste from industrial and commercial establishments, refuse from institutions, market waste, yard waste, street sweepings, and so on. Solid waste management involves the control of waste generation, and waste storage collection, transportation, processing and final disposal in a manner that is in accordance with the best principles of public health, and economic, engineering, aesthetic and environmental consideration. Faced with rapid urbanisation and limited resources, much of the solid waste generated is not collected by the public authorities. Where the waste is collected, it is in most cases disposed of haphazardly by way of crude dumping. Collection is usually confined to a few areas, mainly city centre and high-income areas, and even here collection is usually irregular.

Table 3.5: *Solid waste generated and collected*

Urban Areas	Population	Waste generated (tons/day)	Waste Collected (tons/day)	%waste collected	Disposal method
Dar es Salaam	3,000,000	2,000	200	10.0	Crude dumping
Mwanza	500,000	210	80	38.0	Crude dumping
Moshi	137,200	93	47	50.5	Crude dumping
Shinyanga	135,740	95	24	25.0	Crude dumping
Tabora	160,000	120	12	10.0	Crude dumping
Bukoba	67,000	24	7.2	30.0	Crude dumping
Arusha	250,000	200	NA	NA	Crude dumping
Tanga	187,000	190	NA	NA	Crude dumping
Mtwara	100,000	50	15	30.0	Crude dumping
Songea	113,000	57	NA	NA	Crude dumping
Musoma	80,000	30	7	23.3	Crude dumping
Lindi	59,000	NA	NA	NA	Crude dumping
Iringa	121,000	36	16	46.0	Crude dumping
Dodoma	300,000	NA	NA	NA	Crude dumping
Kigoma	105,000	60	10	17.0	Crude dumping
Morogoro	206,000	260	26	10.0	Crude dumping

Source: Returns from various municipalities compiled in, Rwegasira, *et. al.* (1996).

The hinging of the solution of the problem of solid waste management on acquiring more capital equipment through the public sector has been found to be unsustainable. Thus, a major initiative being undertaken in Dar es Salaam and in other municipalities is privatisation of solid waste management. All indications show that privatisation can be a success; but major issues related to governance especially community participation, transparency, and accountability must be addressed. The role of the various sections of the community, especially women and youth groups in urban solid waste management need to be investigated.

Urban Sanitation

High density living in urban areas necessitates advanced sanitation arrangements to minimise the risk of epidemics. The sanitation situation in many urban areas in Tanzania leaves a lot to be desired. This is especially so in the densely occupied areas in unplanned settlements. Only a minority of urbanites is connected to central sewer networks or to septic tanks. By far, the majority rely on pit latrines. In Dar es Salaam, less than 5% of the City's population is connected to a central sewer network. Of the remaining population, 80% depend on on-site sanitation arrangements (70% pit latrines, 20% septic tanks) and the remaining 20% lack even elementary sanitation facilities. In the municipality of Morogoro 21.3% of the households use water-run toilet facilities, while 82.8% depend on pit latrines (Tanzania, 1998a). In the municipality of Tabora, with 17,565 households, 89% used ordinary pit latrines, 0.7% used VIPs, 4.3% used septic tank and 5.9% had no toilet (Tanzania 1998b). In Bukoba Town, with a population of 46,055 people, 18% use flush toilets, 78% use pit latrines and 4% have no toilet (United Republic of Tanzania, 1998c).

A study carried out in Dar es Salaam found that 13% of the households used flush toilets inside or outside their houses. These were mainly found in high and middle income areas. The rest (87%) used pit latrines (Mwandosya and Meena, 1998). Many of the pit latrines are simple, sometimes makeshift. Mwandosya and Meena (1998) found that 59% of the pit latrines in Dar es Salaam had no roofs, 94% had no ventilation pipes, 34% had no doors and 57% had no pit cover. Many are shared and overused and tend to be filled up quickly particularly in areas with a high water table, or during the rainy season. It is difficult to empty them since they are not accessible by motor traffic; although there are local ways of emptying filled up pit latrines (known as *kutapisha* literally meaning "to cause to vomit"). Emptying is usually done into a pit dug nearby, a process which can be difficult in densely built up areas.

Generally speaking, the government's attitude towards pit latrines has been lukewarm, and there has been a lingering belief in the superiority of water-borne sanitation systems and the possibility of many of the urbanites getting hooked on these. On the other hand, in realisation of the impossibility of realisation of such a belief, there have been some limited efforts to popularise the VIPs, which emit no smell and do not harbour flies. Limited initiatives have been undertaken especially by NGOs to popularise manual pit emptying systems (MAPET)

Nevertheless, the sanitation situation in many urban areas in Tanzania is poor and this is evidenced by the outbreak of water borne diseases such as cholera now and again. It is unlikely that water-borne systems of sanitation will be accessible to the majority of urbanites in the foreseeable future. A recent consultants' report points to the continued relative importance of on site methods of sanitation especially the pit latrines, and the falling relative importance of sewers (Howard and Humphreys, 1995). Research into the technical, legal, cultural and planning aspects of popularising advanced forms of pit latrines is required.

Public Toilets

The busy urban life entails spending considerable time away from their homes and requires that there be public toilets. In countries like England, public toilets and baths were developed to fill the gap resulting from much of private housing not having toilets. The colonial government in Tanzania did provide some public toilets and many of these were run by local authorities. Nevertheless, very little has been done to construct more public toilets or to sanitarily run those that are in existence.

Officially there are 42 public toilets in Dar es Salaam most of them were built before 1970. Most are rundown, boarded and not working. The public is therefore forced to use the bush, or to go to nearby offices, hotels, or institutions (including mosques) to answer the call of nature. The non-existence or poor and filthy public toilets is also evident in areas where normally one would find public toilets such as markets, bus or railways stations and primary schools. Provision of public toilets is wanting because of lack of, or poor, supportive infrastructure such as water, lack of maintenance, overuse and abuse, and lack of resources.

Throughout the 1980s and early 1990s, it was obvious that the DCC had failed to run its 42 public toilets. From 1995, a policy to privatise public toilets in the City was adopted, perhaps taking a cue from Uganda where public toilets are privatised and are working well. The private operators in Dar es Salaam charge between T.shs.50-100 a visit and depending on the service required; and pay a monthly fee of T.shs.5,000 to the DCC. A recent research found out that 94% of those interviewed complained that the existing toilets in the city were too few and sited too far apart, and that they cannot provide adequate services even if they were to be rehabilitated (SDP 1998). Respondents proposed that public toilets should also provide other services such as baths, laundry, hair dressers, barber shops, and luggage stores. This means that the whole role of public toilets could be included in a wider set of urban services.

It is significant that the study found that the majority of the residents in unplanned areas where the provision of private toilets is poor are in favour of public toilets. Some 60.1% proposed that public toilets with bathrooms should be provided in unplanned areas. On whether such services should be paid for, 72.7% said that a service charge should be paid, while 6.7% proposed that public toilets should be managed by communities or street organisations and should be provided free.

It is thus clear that public authorities have the duty to respond to this demand for improved public sanitation and convenience services. It is understood that the DCC is working on bye-laws to privatise public toilets in a transparent manner. The situation of public toilets remains critical in many urban areas in Tanzania. Research is required into how these can be operationalised, taking into account social and cultural attitudes, inadequate water supply, modes of financing and maintenance and so on.

URBAN SOCIAL SERVICES: THE CASE OF PRIMARY SCHOOLS

With increased urbanisation, there has been an enhanced demand for social services such as health facilities and schools. The demand for primary schools in urban areas has increased not only as result of the growing importance of natural increase in urbanisation but also as a result of the realised superiority of urban compared to rural schools, a fact which has encouraged many urbanites to take their kith and kin to study in urban areas.

Despite this increased demand and despite the government's own policy of insisting on universal primary education for every child, investment in primary education has not been commensurate with demand in a number of respects. Although the quality of primary schools in urban areas is perhaps better than in rural areas, urban primary schools are facing a number of problems. Taking the case of Dar es Salaam as an example, the DCC has 167 primary schools but the required number is considered to be between 400-500. This has led to overcrowding in most primary schools. Instead of the normal figure of 45 pupils per class, numbers hover around 100 and go up to 300 in some cases. Many children cannot get admission to primary schools thus increasing the level of illiteracy. Of the required 77,282 desks, only 38,510 (or 49.8%) were available. Many pupils have therefore to sit on the floor (Kironde, 1997b). Accommodation for teachers is also very poor.

Because of over-use and poor maintenance, school buildings are in a poor state; and most schools miss such services as electricity, running water, or toilets. In terms of sanitation, and in the absence of reliable water supply, pit latrines have been constructed in many primary schools in Dar es Salaam. The required number of latrines is 9,350. The available number is only 1,402 (15% of the required number). This means that 176 children are forced to use one latrine instead of the standard 25.

By contrast, according to the data collected from Kisarawe (population 13,000), the township has an excess of desks, classrooms have a maximum of 44 pupils, and the number of pupils per pit latrine is only 20. As it is safe to assume that there is sufficient pressure on parents to ensure that their children go to school, the above situation in Kisarawe reflects the fact that there are fewer children in this district town compared to the city of Dar es Salaam.

Although large urban areas like Dar es Salaam suffer a shortage of teachers, this shortage is worse in rural and in small urban areas, in both quantity and quality. Also access to educational materials is better in larger urban areas.

Urban authorities act as agents for central government in the provision of primary education, but the central government has been under-funding local authorities in terms of education since the 1970s. Faced with limited resources, the government has responded to increased demand for primary education by instituting a number of charges that parents are required to pay. The charges for Mwanza and Dar es Salaam are shown in Table 3.6.

Table 3.6: *Parent's contributions (T shs) towards primary education, Dar es Salaam and Mwanza 1996*

Category of expenditure	*Dar es Salaam*	*Mwanza*
UPE* contribution	1,000.00	1,000.00
Desk contribution	300.00	500.00
Games and sports	200.00	200.00
Mzimuni Education Fund	100.00	--
Buildings	10,000.00	--
School Fencing	500.00**	--
Examinations	50.00	--
Identity Card	--	300.00
Uniform	--	4,700.00
Total	12,250.00	6,700.00

* Universal Primary Education
** Minimum
Source: Kironde, (1997b)

Primary school authorities have faced problems in collecting these charges since government policy has been lenient on defaulting parents insisting that no child should be denied education on account of the parents' failure to meet these charges. Also, for contributions whose total exceeds 5,000/=, permission from the Ministry of Education must be sought.

The quality of primary education has gone down nation-wide, but there is some qualitative difference between the situation in large urban areas and in rural and small urban areas. Children from poor families suffer most. The decline in education must be having some bearing to the quality of the future citizens. There is definitely a need for the government to adopt a deliberate policy of improving education. Parents' contributions should be solicited for and properly utilised. The contribution of the private sector, religious groups and NGOs should be recognised and harnessed. The quality of primary education is likely to improve if local governments are strengthened. The DCC for example, as a result of improving its revenue collection since the abolition of the City Council, has been able to commit some funds to rehabilitation of schools, constructing more classrooms, and contributing to the desk fund. This indicates that the possibility of improving the quality of primary education is there.

URBAN WATER SUPPLY

Rapid urbanisation has increased several fold the demand for water for general use, sanitation, and economic activities such as industries. The demand has in most cases outstripped supply. Supply has been faced with ageing and poorly maintained plant, as well as inadequate and unreliable water sources (which for example get dry or very low during dry seasons). Other problems include, unreliable power supply, dilapidated water distribution systems characterised by leakage, poor record of water users, low tariffs which are moreover poorly collected resulting into low financial capacity, illegal connections, political interference especially with regard to the level of tariffs and disconnection, poor metering, poor customer data, overcentralised decision making and so on. The urban water supply regime in Tanzania is characterised by financial and management problems. The revenue collected is usually far below the costs involved in the supply of water. For example, according to the Director of Mtwara Urban Water and Sanitation Authority, revenue collected from water transactions in Mtwara is Tshs 100m while expenses amount to Tshs 375m. The balance is usually met by the central government, which, for example pays power bills and salaries for permanent employees. Another example is the situation in Dar es Salaam where the average cost of producing 4.5 cubic metres of water between 1990-1995 varied between T.shs. 196 and 557. Yet, by contrast, the domestic tariff varied between T.shs. 57 and 160 (Mwilawi, 1996).

The daily water demand in the 62 registered urban centres in Tanzania is 912,3000m^3 while the average supply is only 494,300m^3. This means only 54% of the demand is actually supplied (Table 3.7). This does not take into account the quality of the water supplied. In many cases, because of malfunctioning water treatment plants and non-availability of water treatment chemicals, or back seepage, water is supplied untreated or partially treated.

Except in a couple of cases, installed capacity is less than demand. Water supplied is also usually lower than water that finally reaches the consumers on account of leakage. In Dar es Salaam for example, the daily demand for water is 90 million litres. Only 60 million litres are pumped and only 40 million litres finally reach the consumer, due to numerous leakages. Use of treated water for irrigation and for the rearing of livestock along the transmission lines, as well as illegal connections, lead to a lot of water getting used without being paid for. According to one study, 53% of the water produced by DAWASA is unaccounted for, and therefore yields no revenue (Howard and Humphreys, 1995).In order to redress the problems of inadequate urban water supply, the government formed a National Urban Water Authority (NUWA) in 1984, to look after the supply of water throughout all urban areas in Tanzania. NUWA was expected to start with Dar es Salaam and gradually extend its jurisdiction to all urban areas in the country. NUWA was never able to move out of Dar es Salaam and parts of the Coastal Region and its management of water supply remained poor. In 1997 NUWA was transformed and restricted to operate in Dar es Salaam; and was further given mandate to take over sewerage services in the city. It was renamed the Dar es Salaam Water and Sewerage Authority (DAWASA).

Water supply in other urban areas was, until recently under the Regional Water Engineer. However,in an attempt to improve operations, the government decided to decentralise water tariffs and formed water boards to run water supplies and sanitation services in Arusha, Moshi and Tanga municipalities. Water and sanitation boards have subsequently been formed for the municipalities of Dodoma and Tabora, Shinyanga, Mwanza, Morogoro, Mbeya, Iringa and Mtwara. The Boards have operational autonomy to set their own water tariffs, and to collect and spend revenue from water sales. Experience in Arusha, Moshi and Tanga water boards has showed encouraging results with water supply improving in terms of quantity, quality, system reliability and revenue collection (Box 3.3).

Table 3.7: *Demand and Supply of Water in Urban Areas in Tanzania*

Region	Daily water demand (m^3)	Daily water Supply (m^3)	Installed Capacity(m^3)	% Demand Supplied
Arusha	42,647	32,541	50,892	76
Dar es Salaam	409,500	191,000	273,000	47
Dodoma	55,870	36,863	43,570	68
Iringa	26,251	13,127	14,772	50
Kagera	13,007	8,380	12,975	64
Kilimanjaro	31,240	16,033	20,722	51
Kigoma	21,095	13,014	14,075	62
Lindi	11,700	4,407	7,600	38
Mara	19,110	9,600	11,300	51
Mbeya	39,555	25,800	31,420	65
Morogoro	38,500	42,647	25,150	62
Mtwara	13,000	7,540	n.a.	58
Mwanza	50,051	43,244	61,244	74
Coast	4,409	3,552	4,680	81
Rukwa	6,770	4,070	4,660	60
Ruvuma	11,070	5,633	5,833	51
Shinyanga	16,751	10,995	27,260	66
Singida	9,041	3,925	3,000	43
Tabora	28,222	10,379	15,417	37
Tanga	56,792	30,525	33,672	54
Total	915,581	494,629	661,288	54

Source: DAWASA and Regional Water Engineers, 1997

It is clear that the government's approach to meeting the demand for urban water has been to seek technical and managerial efficiency within a framework where urban water supply and management is placed in the hands of central government agencies. Issues that need considering for both research and policy initiatives include the following:

- The role which local governments should play in urban water supply and management. There has been a number of calls to let local governments handle water supply on the argument that water is a very crucial item and should be supplied by an organ that is responsible to the local population. The counter argument has been that urban water supply is such a complicated undertaking which moreover requires huge investment and sophisticated co-ordination, and that local governments do not have the necessary capacity to deliver;

- It is unlikely that the conventional ways of water supply and management will meet the demand in the foreseeable future. There is need to look into alternative ways of meeting demand; like using water from local sources (wells, bore-holes, rivers, rain water and so on) and

BOX 3.3: SOME EXPERIENCES FROM THE ARUSHA AND TANGA URBAN WATER SUPPLY AND SEWERAGE BOARDS

The Urban Water Supply and Sewerage Boards of Tanga and Arusha came into operation in July 1994. The quality of service and revenue collection has improved since then. The daily water demand in the municipalities of Arusha and Tanga is 42,000m³ and 46,000m³ respectively. In 1994, the supply was 30,000m³ and 21,600m³ but had in 1997 gone up to 35,000m³ and 25,000m³ for Arusha and Tanga respectively. This is still below demand and rationing is practised. Customer data is being improved by doing a house to house survey. Non availability of maps and the unplanned nature of many areas are two major problems. In Tanga, it was observed that in the supply area of 18,300 premises, only 10,900 customers were connected to the water. The rest were getting water from the premises that are connected, and, in the absence of metering, they were not paying. Non metering has been found to lead to water wastage and a metering campaign is being implemented. So far, 25% and 50% of the customers in Tanga and Arusha respectively are metered. Water tariffs have been revised upwards from an average of Tshs 35 per cubic meter to Tshs 187 in Tanga; and from Tshs 35 per cubic metre to Tshs 174 in Arusha, in 1994 and 1997 respectively. This is still below what water vendors charge which is between Tshs. 250 and 500 a cubic meter. The minimum charge has also gone up from Tshs 350 in 1994 to Tshs 1,500 in 1997 in Tanga, and from Tshs 800 in 1995 to Tshs 3,000 in 1998 in Arusha. Debt collection has been enhanced although the most indebted customers are government institutions. Major cost centres, that is, electricity, water treatment chemicals and personnel are being reviewed with a view of cutting down unnecessary costs. As a result of the above measures, revenue collection has improved and for the past three years, revenue has exceeded cost. Operation and maintenance costs and some limited investment have been financed from own sources. Only the bill for salaries for permanent staff has been borne by the central government.

Source: Munisi, A.A. and Mutegeki, M.T.K. "The experience, challenges and opportunities in financing operations and maintenance of urban water supply." Paper presented at the Urban Sector Rehabilitation Launch Project, Bagamoyo, 11-12 November, 1997.

purifying it locally; as well as locally managed water distribution systems. An interesting example worth following closely is the community water supply being instituted in Tabata and Kijitonyama, completely managed by local CBOs;

- The extent to which privatisation can help in improving urban water supply and management needs to be looked into. Some privatisation has been adopted in neighbouring Kenya and Uganda (UNHCS, 1998). Consideration of privatisation in Tanzania is still limited to discussion on privatising parastatal organisations like DAWASA, but the whole array of privatisation options can be invoked to improve urban water supply;
- In terms of poverty and social equity, it is noticeable that much of the current water supply is subsidised, yet the beneficiaries are those households living in high income areas. Most residents in unplanned areas are not connected to mains water supply; in many cases they have to depend on water sellers whereby they pay more for water consumption than those who get supply from the public sector (see Box 3.3). The question of continued direct or indirect subsidisation of water needs to be looked into.

URBAN ENERGY SUPPLY

Energy supply is critical to urban areas in order to support economic activities such as industries, public utilities such as water supply, social infrastructure such as schools, and hospitals, and to provide lighting and power for domestic, commercial and public use.

Regional and some district centres in Tanzania have some form of electricity supply either from the national grid system or from local thermal generator suppliers. Nevertheless many district towns and most of the townships, such as Ngara, Ukerewe, Sengerema, and Geita, are not supplied with electricity. Besides, for those urban areas which are supplied with electricity, the supply is usually below demand, notwithstanding the fact that many households and firms in these urban areas are not connected to electricity. There are areas such as Arusha, where the power supply exceeds demand but these are exceptions rather than the rule. Whether the power supplied meets demand or not, most households in urban areas are not connected to electricity as figures shown in Table 3.8 indicate.

Table 3.8: *Households supplied with electricity in 1988 in some selected municipalities*

District	Total number of households	% Households supplied with electricity
Bukoba Urban	10,147	26.7
Morogoro Urban	26,702	40.4
Tabora Urban	17,565	31.6
Dar es Salaam	—	34.3

Source: Population Census, 1988

Many households in urban areas depend on paraffin for lighting, and paraffin, charcoal or firewood for cooking. This has implications for pollution and national environmental protection. In addition, all urban areas suffer from constant power interruptions as a result of inadequate supply due to technical problems, ageing machinery, lack of fuel or spare parts, low water levels especially during the dry season, and so on. The government's policy towards urban power demand and supply has been to rely on a national parastatal, TANESCO, to be the sole supplier of electricity in the country. TANESCO has been beset by many managerial and technical problems including lengthy and bureaucratic response to consumer demand for connection or fault remedying, inefficient meter recording, billing, and collection, low tariffs, a poor investment (expansion) policy, a poor maintenance policy, and so on. Movement towards privatisation is still unclear although a private power supply company was allowed to generate electricity and sell it to TANESCO. This however, has not come into operation because of disagreements between the private company and the government about the level of tariffs to be charged. The move by TANESCO to introduce pre-payment systems (LUKU) proved popular but has (surprisingly) not been universally introduced. The government and public corporations remain major debtors to the electricity supply company.

Research and policy initiatives are required on the economics and feasibility of privatisation of power generation and distribution in the country. The fact that the majority of urbanites are not connected to electricity has connotations for urban poverty since it may well be that the poor who are not connected pay more in paraffin, firewood and charcoal bills compared to those who are connected to electricity. There is need to increase competition in the national power supply system and do away with TANESCO's monopoly which has been in existence for decades. Like is the case with water supply, the role of local authorities in managing and distributing local power supply need to be investigated, as well as the possibilities of having small or medium scale power suppliers, as alternatives or supplementary to nation-wide suppliers.

URBAN TRANSPORTATION

The rapidly growing population needs to travel often particularly since in most instances, the place of work and place of residence are usually far apart. Travel is also required to markets, social and cultural activities, hospitals and so on. The need for organised public transport began to be felt in Dar es Salaam in the 1940s and for many years it was only the City of Dar es Salaam, which had public transport in the country. Today, many other urban areas have their forms of public transport, much of it privately owned. Providing urban public transport through the public sector was the norm in many urban areas in Africa until relatively recently.

The private company that was running public transport in Dar es Salaam since 1949 (Dar es Salaam Motor Transport -DMT) was nationalised in 1970 and was renamed UDA (Usafiri Dar es Salaam) in 1974. The government policy was to protect the monopoly of UDA in providing public transport services although during the latter part of the 1970s it was clear that UDA was not managing. Buses were infrequent as well as overcrowded. Private pirate transporters who set up business and whose buses were known as *thumni thumni* were banned by the government. However, the situation got worse and in 1983 the government legalised private urban transporters in Dar es Salaam, known as *daladala*. Initially *daladala* operators were required to pay part of their monthly earnings to UDA. Nevertheless, the performance of UDA has deteriorated almost to non-existence as the available figures for 1984-1993 show (Table 3.9)

Table 3.9: *UDA performance, 1984-1993*

Year	Buses owned	Buses on the road	Passengers ferried (m.)
1984	216	141	100.4
1985	216	139	97.8
1986	216	108	80.6
1987	223	101	74.7
1988	204	109	73.4
1989	172	70	51.9
1990	166	69	31.3
1991	94	32	19.6
1992	77	25	11.2
1993	87	36	17.6

Source: UDA Records.

In 1996 UDA was making an average monthly loss of Tshs 17.3m, had 33 buses on full time and only 20 buses worked the whole day without breakdown. Currently, UDA has only 16 buses and running only 4 routes. UDA's problems include ageing buses, lack of spare parts, stiff competition, lack of market research, reduced competitive spirit among employees and so on.

Public transport in Dar es Salaam and in other large towns is provided almost entirely by the private sector. In 1997 according to the *daladala* operators organisation (MUWADA) some 7654 *daladala's* were registered in Dar es Salaam. To these must be added many others which operate as pirates, without licence. These private transporters have showed a great degree of flexibility, ingenuity and adaptability. Observations in Dar es Salaam indicate that the transporters started by using large rundown buses taken from long-distance routes. They then used converted pick-ups (which came to be known as *chai maharage)*. These have now largely been superseded by mini-buses. These transporters have shown the ability to venture into new routes even when the roads are bad; and to suit types of vehicles to road conditions. Landrovers for example are used for new, poorly serviced areas, like Tabata and Kisukuru in Dar es Salaam. Competition is keen and transporters have been forced to improve the quality of the buses, and to be flexible with the fares.

While the policy of liberalising the transport sector is commendable, the government is yet to adopt a clear supportive role towards private transporters. There is little doubt that these need to be regulated. The regulations must not be oppressive but must be the minimum required to ensure safety, and law and order. The government's role must also be to improve infrastructure such as roads, bus stops, lay-by's and so on. Attempts to control fares in Dar es Salaam were discredited when, in 1996 the government put up fares from T.shs.100 to 150. The public protested since there was no justification for such an increase. After a short stint characterised by boycotts and angry exchanges between the public and the bus operators, the latter abandoned their pressing for higher fares. There has been stability in the fares for some two years now. Good relations between the Association of Bus Operators and the government can go a long way towards addressing some of the problems that are cropping up. The public for example are complaining of abusive language and rowdy behaviour by touts, over-speeding, random stopping, indecent dressing by operators, discriminating passengers (especially the rejection of school children), lawlessness on the roads, non-completion of routes and so on. These are some of the areas where positive public intervention can be useful.

The question of urban public transport is complex and must be addressed in broader terms than buses especially in large cities like Dar es Salaam. Non-motorised transport must be addressed. Many people, especially the poor, walk to their workplaces. Many others use bicycles but these are not catered for. Most roads are not provided with separate pedestrian or bicycle routes which moreover could be lined with shade trees. The Sustainable Dar es Salaam Project attempted to address the issue of non-motorised urban transport but these attempts did not go far. This is an area where research could be useful. Decentralisation of urban activities is another area, which need to be considered. Smaller urban areas which do not feel the problem of public transport at the moment could well be advised to ensure that as their urban areas grow, activities are also spread and do not get concentrated in the central areas.

URBAN SECURITY AND ACCESS TO JUSTICE

Almost everywhere in Africa, urban areas are becoming very insecure. A multitude of reasons are creating a situation which has resulted into an increased rate of muggings, armed robbery, rape, street justice, destruction of property and so on. Realising the problem an African Forum on Urban Security was launched in Dakar, Senegal in February 1998. In Dar es Salaam in August 1998, a "Safer Dar es Salaam Project" was launched. At that launching, it was learnt from the Ministry of Home Affairs that some 26% of all the crimes reported to the police nation-wide were committed in Dar es Salaam. Increased fencing and use of grills in buildings, and the hiring of private guards demonstrate magnitude of urban insecurity. Urban residents in major areas, unlike residents in rural areas and in small towns need increased security. Governments are being enjoined to increase urban security.

The question of security is serious in urban areas because of the high numbers of population; the varied nature of peoples' backgrounds; the high demands of urban life coupled with limited income generating activities; and the poor level of individual identification. The situation is made worse in cities like Dar es Salaam where the majority live in unplanned areas; where there is no national or local personal registration or identification systems; where most streets and houses are not named or numbered; where ward and neighbourhood governments are non-existent or very weak; and where the police are few, poorly paid and ill-equipped. Information collected for this study from Magomeni Police Station, incorporating reports from the police posts of Manzese, Mabibo, Urafiki, Kimara and Magomeni itself for the period 1987-

1997, indicate an alarming urban security problem for Dar es Salaam. Nearly 300,000 incidents of crime were reported including instant justice (killing), rape, mugging, sodomy and defilement. If this is extrapolated to cover the whole city, it is possible that one is talking of over 2 million incidences of crime over the period. The highest incidence of crime was mugging; although it must be noted that many more cases are not reported to the police for one reason or another.

Public reaction to increased insecurity has been to institute instant justice (usually stoning or burning to death) particularly to suspected thieves on the argument that the police and the courts tend to let criminals walk away free only to continue with their heinous acts. Others feel that instant justice is the best deterrent.

One approach of the government to increased urban insecurity was to institute compulsory urban vigilante groups (*sungusungu)* in the early 1990s. This was received with mixed reactions, with some arguing strongly that security was the job of the police. Vigilante groups have diminished in their activities although they are still active in a number of urban areas where the residents are willing to undertake them.

Initiatives are urgently required to improve urban security. These definitely include improved local governance. In the single party era, the party ten-cell leader was an important institution in ensuring local governance. Ten cell leaders no longer have government powers, and plans to strengthen local government at ward, and neighbourhood levels are going on at a very slow pace. Improved information systems on property and urban residents including compulsory registration with local government and the introduction of identity cards could help somewhat. Deterrent measures such as increasing the efficiency of the police are called for as is the increased role of the public in ensuring security at local level in a lawful manner. Improved access to income, services, and social security are also called for.

ASPECTS OF GOVERNANCE FOR URBAN DEVELOPMENT

It has been argued in this chapter that the government has been unable to address many of the demands of rapid urbanisation. One factor bringing about this situation is the weakness of local governments in the country. This weakness has a number of facets. One, many local governments do not raise sufficient revenue to effectively manage urban development, and have thus to rely on central government grants. Between 1984-1996 local authorities in Tanzania depended on central government allocations for between 55-70% of their revenue.

Two, is the question of poor conceptualisation of local governments where both the officials and the peoples' representatives (councillors), have regarded themselves as agents of the central government, who were in local government to maximise their interests rather than the interests of their citizens. Many local authorities are facing a managerial problem or crisis of legitimacy.

Three, in terms democracy, local authorities are very weak. The level of accountability and transparency to the local citizenry leaves a lot to be desired. "Tax revolts" have been recorded in a number of localities such as Arumeru, Morogoro, Tanga and Korogwe, tax payers demanding to be told what their money is being used for, but officials and councillors are reluctant to do so.

The weaknesses of local governments has been documented countless times. The ongoing Local Government Reform Agenda aims at creating democratic and autonomous local authorities which are politically accountable and are transparent. Autonomy involves the ability to be able to make many decisions without having to look over the shoulder, and also to be financially sound. The principle of subsidiarity is being adopted aimed at allocating roles to the lowest levels deemed best for execution and which enable public participation in setting priorities. Cost effectiveness in the provision of services is being encouraged. Lastly, local authorities are expected to engage in poverty reduction activities through economic development, by, among others, fostering partnerships with the private sector and civil society institutions. These measures are currently being put in place. Results are thus yet to be seen.

An effective government policy to respond to the demands of rapid urbanisation is greatly hinged on the extent to which local governments from the grass-root level can be enhanced. Although there are steps in that direction, the current pace is too slow when, in the meanwhile urbanisation is going on rapidly without proper management. Some of these steps are inadequate. For example, in cases of land management and the provision of public utility services there are lingering steps towards centralisation, when what is required is effective decentralisation.

CONCLUSIONS

This chapter has surveyed a number of areas where rapid urbanisation has generated demand, and has looked at the steps which governments, local or national have taken to address these demands. Most of these responses have not been able to cope with the demand. In all cases, it appears that the question of improved governance is of crucial importance.

This revolves around strengthening local governments as well as the civil society, and also incorporating more actors. Such actors include the private sector, the donor community as well as community-based organisations and institutions. In each area of policy surveyed, established standards seem to be overtaken by the mere volume of urbanisation. New forms of coping strategies are called for to deal with the problems. There is therefore a lot to learn both on the part of the government and policy makers and on the part of researchers so that suitable forms of intervention are formulated, and the role of government which in most cases would be supportive and regulatory, is defined.

REFERENCES

Becker, C.M., Hamer, A.M., and Morrison, A.R. (1994)
Beyond Urban Bias in Africa: Urbanisation in an Era of Structural Adjustment, James Currey, London and Portsmouth.

Confederation of Tanzania Industries (CTI) (1997)
"A Study of the Dar es Salaam City Commission Finances," Economic and Social Research Foundation, Dar es Salaam.

Howard Humphreys (1995)
"Rehabilitation of Dar es Salaam Water Supply System: Main Report," Prepared for NUWA, Dar es Salaam.

Kabwogi, C. (1997)
"Rent Restriction in Tanzania," *Phd* Thesis, University of Sheffield.

Kironde, J.M.L. (1996)
"The Role of Municipalities in Dealing With Urban Poverty in Tanzania: The Case of the City of Dar es Salaam." Report Prepared for the UNCHS-Habitat on Urban Poverty.

Kironde, J.M.L. (1997a)
"Land Policy Options for Urban Tanzania," *Land Use Policy*, Vol. 14, No. 2, p. 99-117.

Kironde, J.M.L. (1997b)
"Financing the Sustainable Development of Cities: The Case of Dar es Salaam and Mwanza in Tanzania." Report prepared for UNCHS-Habitat, Nairobi.

Kironde, J.M.L. (1998)
"The New National Land Policy and its Possible Impacts on Urban Land Management and Development in Tanzania," Paper presented at the international workshop on 'Comparative Policy Perspectives on Urban Land Market Reform in Latin America, Southern Africa, and Eastern Europe' held at the Lincoln Institute of Land Policy, Cambridge Mass, 7-9 July.

Kombe, W.J. (1994)
"The demise of Public urban land management and the emergence of the informal land markets in Tanzania: A Case of Dar es Salaam City", *Habitat International*, Vol. 18, No. 1, p. 23-43.

Kombe, W. (1997)
"Regularizing Urban Land Development During the Transition to Market-led Land Supply in Tanzania." in, W.J. Kombe and V. Kreibich (eds), *Urban Land Management and the*

Transition to a Market Economy in Tanzania," Spring Research Series No. 19, University of Dortmund, p. 29-47.

Mabogunje, A.L. (1990)
"Urban Planning and the Post-colonial State," *African Studies Review,* Vol. 33, No. 2, p. 121-203.

Magoho, A. (1991)
"State and Urban Unemployment in Tanzania: An Appraisal of the Human Resources Deployment Act." *LLM* Thesis, University of Dar es Salaam.

Mwamfupe, D. G. (1994)
"Changes in Agricultural Land Use in the Periurban Zones of Dar es Salaam, Tanzania." *PhD* Thesis, University of Glasgow.

Mwandosya, M. J. and Meena, H.E. (1998)
Dar es Salaam Water Demand: An End User Perspective, Centre for Energy, Environment, Science and Technology.

Mwilawi, D.R. (1996)
"The Problem of Cost Recovery of Public Water Utilities inTanzania: The Case of NUWA." *MBA* Dissertation, University of Dar es Salaam.

Pugh, C. (1992)
"Land Policies and Low Income Housing in Developing Countries: A Review With Reference to Kenya and India," *Land Use Policy,* Vol. 9, No. 1, p. 47-63.

Rwegasira, M.B., Chaggu, E.J., Kasseva, M.E., and Kassenga, G.R., (eds) (1996)
Proceedings and Papers from the Training Workshop on Solid Waste Management for Officers Responsible for Municipal Solid Waste Management, held in Dodoma, 25-30 November, 1996, UCLAS and NEMC, Dar es Salaam.

S. D. P, (1998)
"Managing Public Toilets: Research Report," Dar es Salaam City Commission.

Stren, R. R. (1992)
"Large Cities in the Third World," in, UNCHS, *Metropolitan Planning and Management in the Developing World,* Nairobi.

Swai, B. O. (1998)
"Urban Poverty Reduction with a focus on employment, incomes and assets; The case of Dar es Salaam City, Tanzania," *Journal of Building and Land Development,* Vol. 5 No. 1, 1998, p. 68-71

U R T, (1991)
The National Informal Sector Survey, Planning Commission, Dar es Salaam.

U R T, (1995)
National Land Policy, Ministry of Lands Housing and Urban Development, Dar es Salaam.

U R T, (1998a)
Morogoro Region Socio-economic Profile, The Planning Commission Dar es Salaam and the Regional Commissioners' Office, Morogoro.

U R T, (1998b)
Tabora Region Socio-economic Profile, The Planning Commission Dar es Salaam and the Regional Commissioners' Office, Tabora.

U R T, (1998c)
Kagera Region Socio-economic Profile, The Planning Commission Dar es Salaam and the Regional Commissioners' Office, Kagera.

U RT, (1998d)
1998/99 Budget Speech Ministry of Lands ands Human Settlements, Government Printer, Dar es Salaam.

UNCHS, (1996)
An Urbanising World: Global Report on Human Settlements, Oxford University Press.

UNCHS, (1998)
"Privatisation of Municipal Services in Eastern Africa: A Governance Approach to Human Settlements Management." Research Report supported by the Ford Foundation, Nairobi.

Wenga, P., Andrea, I. and Schulz, M. (1995)
"Self-help Organisations in the Informal Sector of the Dar es Salaam Region." A Study Report prepared for the ILO.

World Bank, (1994)
World Development Report 1994: Infrastructure for Development, Oxford University Press, New York.

4

ASPECTS OF MIGRATION AND URBANISATION IN TANZANIA

*Adolfo Mascarenhas**

INTRODUCTION

Any serious study of migration and urbanisation in Tanzania reveals some of the deep-seated contradictions in the initial and subsequent attempts to control the processes at work.[1] The national psyche was that towns and cities were alien to the country,[2] full of foreigners with corrupting and capitalistic tendencies. Those living in cities were regarded as 'migrants' who would return to Asia or Europe. The major contradiction is that the judicial principles of the colonial native authority system, under new names, are sub-consciously retained. As a result even indigenous peoples are regarded as temporary residents and periodically, they can be rounded up and "deported" to their rural home areas.

The spectacular growth of the urban population, especially that of Dar es Salaam, which has been revealed every decade, (after the national censuses), indicates that there are profound changes taking place in Tanzania. The only official acknowledgement that major changes were taking place was through the censuses. The national censuses have in a number of ways been one of the main source of information on, urbanisation, migration and the general status of the population of Tanzania. However, it should be noted that while the first two censuses made much progress, the 1988 census was analysed less rigorously and indeed the subject of urbanisation was almost left out. The much anticipated fourth national census which was to take place in August 1998, has for a number of reasons been postponed. Therefore at a very critical decade we have very few hard figures about the level of urbanisation.

* Research Professor, Institute of Resource Assessment, University of Dar es Salaam.

However, it should be emphasised here that the censuses tell us very little about the processes at work. For instance, the national census does not reveal the forced migration initiated by the government and the Party in the mid-1970s. Most of the movements during villagisation were for short distances of about 3-5 kms. This major upheaval is unrecorded by the censuses.

The problem about urbanisation and migration is not only about some of the anomalies that exist and the rapid population growth, but also the processes at work. The processes at work are not static but can, and have altered radically. For instance, the 1990's have seen more female migrants than ever before. As more and more people enter into the large urban areas, livelihood in them is becoming more precarious because of greed, corruption, desperation and sheer manipulation by the rich and the powerful. It is here that there is need to pay attention to the issue of governance as well as a new research agenda.

URBANISATION AND MIGRATION: AN OVERVIEW

Urbanisation

Both urbanisation and migration are very complex manifestations that go beyond the census figures. While census figures periodically tell us whether growth or stagnation is taking place, they tell us little about why or what or which process or activity is taking place. Even the basic term "urbanisation" or its derivations can take different meanings (eg persons living in urban areas, the social process of adjustment, the physical spread of cities etc, (Tannerfeldt, G. 1995). In Africa, because of the great disparity between urban and rural areas, one is struck by the links between urbanisation and development.

Ultimately development is about improving the quality of life for all people. There are various schools which have different or overlapping components which go into consideration. These are administrative experience, sustainability' balance, self-reliance, equity and empowerment.[3] Within this context there are two very fundamentally opposing views about urbanisation; those who in some form or another believe that cities are the motor to development and those who view them as parasitic.

Theories about the growth of towns and cities abound. From a planning point of view, the internal configuration of the urban areas, especially for the larger centres, was strongly influenced by the British Town and Country Planning Ordinance of the late 1940's. It was replicated, used and influenced the Town Planning Department in Tanganyika and Tanzania for several decades. The broader picture emerged with Tanzania's early exposure to regional planners from France; and a start could be made from that perspective.

Perroux's concept of "growth poles" expounded in 1949, can be regarded as economic space with its field of forces "from which centrifugal forces emanate and to which centripetal forces are attracted." Each centre has its attraction and repulsion, in which the scale of operation, dominance and surpluses to innovate and to propel can be identified. Propulsive industries are fast growing sectors. Growth poles can be given a spatial dimension in which space can be regarded as "homogeneous" (such as in uniform regions), "polarized," interdependent between economic units, or are hierarchial; or since development does not occur everywhere but at points, growth poles can also be used as a planning tool.

As argued by Doherty (1976) more than two decades ago, growth poles and growth centres would have relevance in development and spatial planning if they were conceived with the aim of: (a) fostering internal spatial integration and lessening external dependency; (b) providing organic links between structure of production and needs of a particular society; (c) establishing forward and backward linkages to the other sectors of the regional and national economy; (d) inducing multiplier effects which initiate growth in the sector producing for the local market; (e) establishing new paths for diffusion of information and innovation; (f) strengthening the local service sector and generating external economy and making the entire economy grow. In this way instead of being alien and parasitic, growth poles could lead to development.

Migration

Regarding migration, the situation is even more complex. For instance, the conventional approaches to the study of migration in Africa have been provocatively criticised and presented in the context of globalisation of the capitalist economy. Migration studies demand more structural, historical and holistic methodologies (Amin, S. 1995). The processes and impacts of migration are linked to a number of economic, social and political policy issues and activities on individuals and groups. Even more complex are the individual decisions, partly based on economic or social factors but also including powerful elements of the "national psyche" best exemplified by the "go west" syndrome found in the United States. Since independence, there has been a powerful "Go-to-Dar es Salaam" syndrome. The reasons for migration have been ignored, signs misinterpreted and the options and possibilities for real participatory development have been wasted.

Some commentators have stated that Africa is a continent perpetually on the move. In a recent well-documented overview on migration it was pointed out that migration and urban studies were first undertaken by sociologists within the perspective of social structures and modernisation. Economists followed because of the pervasiveness of the different types of labour migration and their scale and dynamics which were in response to rural-urban differentials. Demographers on the other hand, focused on the differentials and selectivity of the migrants and considerable progress was made especially in the 1960's and 1970's.

In a review of over 1500 studies on migration in Africa, it was found that about a third were generally on inter-relationships and policies, followed by about 14% each on urbanisation and labour migration. In contrast there was a paucity of material on such topics as female migration (2.5%); migration and regional integration (2.3%) and nomadism (1.7%) (Adepoju, 1995). There were also very few studies even on refugees (6.8%), but with the recent spate of studies, this in all probability, has more than doubled. The quality of data continues to be elusive and some of the gaps are persistent such as those associated with gender issues, are persistent.

An aspect that is important in our understanding of migration to the urban area is the role of the "environment." In this context "migration represents one important demographic response to conditions of poverty and environmental stress in Africa (Krokfors, 1995). Unfortunately little work has been done in this field.

Given this background and the exploratory nature of this chapter, there are bound to be frustrating gaps. Local and national experiences on how migration has an impact on development and how development has an impact on migration have not been undertaken systematically. The lack of migration surveys and the heavy reliance on census information have added to the problems of synthesis (Gould, 1995).

However, since both topics have been generally neglected, it is far better to do something now and to at least have a benchmark of some sort. It is worth briefly to examine what type of information is available for Tanzania.

Bibliographic Sources

There are four significant bibliographic sources of information now available on migration in Tanzania (Campbell and Pons, 1987; Mascarenhas, 1993; Demographic Unit,1997; and Mbonile, 1997). In a general way, mention must also be made of two collections of research pertaining to Africa: these are the work on the rural-urban interface in

Africa (Baker and Pedersen, 1992) and the migration experience in Africa (Baker and Aina, 1995).

The comprehensive bibliography on population and development in Tanzania by Mascarenhas (1993), is divided into 13 sections: sections VI & VII with 26 & 34 items are specifically devoted to "Migration and its Consequences," and "Urbanisation," respectively. Its main value is the references to some valuable research on employment, some socio-economic research on squatter areas of Dar es Salaam etc; but above all it analyses the main trends in migration documentation. Among the conclusions, four are of special importance: first, that there was a bias towards studies on the rural urban migration, secondly, that there were more studies generated from the 1967 census than in subsequent ones and that while more data was gathered in subsequent censuses there was less analysis; thirdly the importance of the non-census data for understanding and analysis such as the National Urban Mobility, Employment and Income Survey of Tanzania (NUMEIST) (Sabot 1974), and other related surveys; fourthly the need to pay more attention to policies and migration trends.

The annotated catalogue of postgraduate thesis of the Demographic Unit (1997) is recent but rather restrictive and only itemises dissertations done in demography, in the mid-1980's when the Unit was formed, at the University of Dar es Salaam. Unfortunately, no effort was made to do a trend analysis of topics or areas covered.

Mbonile's work on 'migration' identifies nearly 106 sources of information. Importantly, because of the sheer volume of forced migration and the resulting massive size of the settlements, it will be appreciated that nearly 40% of the citations are on the refugee influx into Tanzania. Campbell and Pon's (1987) work is a valuable source of information on urbanisation available outside Tanzania. It was not possible, within the short time for this review to get access to other known bibliographic sources on urbanisation.

Census Information

In Tanzania, for a number of historical, economic, and political reasons urban areas were regarded as parasitic. Town councils were abolished in the early post independence years by the post-conial state; and even though they were formally re-instituted in 1982, a great deal of ambivalence still exisists. Therefore, the attitude in Tanzania towards "urbanisation" as reflected by Dar es Salaam, has been mostly schizoid. Urbanisation was the antithesis to the purity and wholesomeness of rural development. It was full of business people that exploited and beggared the rural people. Legally, there was almost no work done to gazette towns and urban places. This means that the census data - one of the major sources of information- is very incomplete.

Hard data and information on migration is of recent origin and is largely based on the various censuses that have taken place. There were several counts during the colonial period. National censuses have taken place in 1967, 1978 and in 1988. The 1998 census was postponed indefinitely.

With very few exceptions in the 1930's and 1950's, migration into urban areas was not regarded as a significant issue. For a considerable period in recent years the population of Tanzania, defined as urban was extremely low. Even by the end of the colonial period only 4% of the population was considered urban. During the first national census - taken six years after independence despite the rapid growth - the urban population for Tanzania was only about six hundred thousands. Only 5.7 % of the total population was regarded as urban (Mascarenhas, 1973). In the 1988 census the urban population had risen to an estimated 4.043 million people or about 17.9 % of the population (URT, 1990; see also Table 4.1 below).

Table 4.I: *Total and Urban Population of Tanzania* (1948-1995)

Year	*Total Population*	*Urban Population*	*% in Urban Areas*
1948	7,744,600	183,862	2.4
1957	9,087,600	364,072	4.0
1967	11,958,654	685,547	5.7
1978	17,036,499	2,265,854	13.3
1988	22,533,758	4,043,684	17.9
1995*	26,800,000	7,487,000	27.9

Source: Various Census Publications. *= Estimated

Unfortunately, the analysis of the section on urbanisation was done in a pedestrian fashion. The census data on urbanisation for Tanzania tends to underestimate the numbers of people but above all, misrepresents the prevailing urban hierarchy in Tanzania. The twenty or so urban areas defined have been the focus of attention and the others have been neglected. Efforts have been made to improve the information available from the census. For instance, during the 1967 census, a total of 204 centres were identified and an estimate made of their population (Mascarenhas, 1973). In 1978, a total of 110 urban centres were identified (Rafiq, 1988).

The current overall pattern of urbanisation in Tanzania is generally not well understood. While the growth of Dar es Salaam draws so much attention, we have little idea about how and why the medium sized towns are growing. The absence of a national policy on development and physical planning will mean wasted opportunities even for other urban areas.

More recently the Ministry of Lands, Housing and Urban Development has tried to take stock of the settlement distribution in Tanzania. Table 4.2, has been reconstituted to give us a composite picture. In a separate exercise the Bureau of Statistics (URT, 1998) identified some 435 small trading centres, in addition to the 26 other urban areas. However, even the 435 centres are probably an under-estimate. One could make a qualified assumption that by the beginning of the year 2000 about one third of the population will be living in urban areas.

Therefore, simply on the proportion of people affected, there are reasons to be concerned and to undertake studies to improve our knowledge about the role and importance of these centres in the development process especially at a local level. There is need for caution. For instance, recently it was found that nearly 30% of the population of Bukoba Town had been excluded from the 1988 census enumeration partly because of commuting patterns (Rwechungura, 1998). The same could also be said of Moshi, Arusha and Mbeya. The implication and impact of such under-estimation requires urgent attention because such information could be used, for instance, to plan for improved services, electricity, water and health services to the rural areas.

"Migration" and "Urbanisation" Issues

The trinity of determinants of population growth or decline is fertility, mortality and migration. Of this trio, migration is the most complex to conceptualize and measure. For instance, while in an absolute sense fertility leads to growth and mortality to decline, by contrast, migration at any time leads to growth in one place and decline in another. Migration is further complicated because it could be a transition stage, periodic, circulatory, temporary or even permanent. A common denominator in migration is mobility or movement, but migration also has both a temporal and a spatial dimension. Social mobility on the other hand, need not lead to physical movement of a great distance. In the social sciences, the literature on migration has to cope with both a physical, biological, and social transaction and "not just an unequivocal biological event" like fertility or mortality (Zelinsky, 1971). Consequently, demographic studies cannot just be concerned with mathematical predictions.

From all the studies it is evident that in Tanzania there has been a propensity for many people to migrate. The basic pattern of migration since independence has altered substantially with movement to Dar es Salaam still being significant but no longer outweighing in importance *via a vis* movements to the other centres. There have been changes and the district and other small towns are having a more dynamic role to the extent that the medium sized towns (not Dar es Salaam) are absorbing most of the population (*see* Table 4.2 below).

Table 4.2: *Type of centre and population*

Type of Town/ Centre	Total Urban Population	Percent of Population	No of Centres	Percent of Centres
Medium	2,143,853	28.28	149	50.7
Large	1,577,258	20.80	56	19.1
City	1,360,850	17.95	1	0.3
Municipality	1,112,017	14.67	7	2.4
Township	893,021	11.78	11	3.7
Small	463,521	6.11	62	21.1
Others	0,143	0.40	8	2.7
Total	7,580,663	100.00	294	100.0

Explanatory Notes: Medium = (10,000 - 19,999); Large Towns = (20,000 - 49,000); City = (Over 1 million); Municipality = (over 100,000); Townships = (50,000 - 99,999); Small = (5,000 - 9,999); Others =(less than 5,000)

Source: Derived from: URT (1994)

Prior to independence in 1961, the majority of migrants moved from the rural areas towards the more economically active regions such as Tanga, Morogoro and Moshi/Arusha. However even by the 1970's conditions had begun to change. In one study, it was found that as many as 9.5% of the rural population of Arusha and 70% of Arusha Town were migrants i.e the regional, *in situ* migration was becoming very important (Mlay, 1981). Short distance migration to other urban and rural areas has probably increased substantially in some parts. For instance, it is hard to explain how population densities in some areas could depart so significantly from the national average increase, unless there was migration. While the percentage increase of density between 1978-88 for Tanzania was 32%; some regions had high figures, notably: Dar es Salaam (77%), Lindi (53%), Mara (49%), Singida (45%).

During the 1978 census an effort was made to find the impact of "migration" on the regional growth of population. This was computed on the basis of deducting the annual growth rate from the natural increase; the difference between the two then gave the gain or loss experienced. Unfortunately, while the 1978 census gave the annual rate of increase by districts, the natural increase was not readily available for the districts. During the 1988 Census, Dar es Salaam continued, as a region, to have a positive migration rate of 5.4% per annum followed by Rukwa and nine other regions. Three of the largest of the nine regional losers were: Coast (-1.2%), Kilimanjaro (-0.6%) and Mara (-0.6%). Mention must also be made of Zanzibar, which overall had a loss of -0.8%. If all relevant calculations had been done at the district level, more refined data would be available for improved analysis of migration. But even then we would still need to have socio-economic information.

Socio-economic Aspects

Since migration is not dependent on demography or censuses but on historical factors - impacts of political, social and economic activities - such factors and studies become important references for understanding migration and urbanisation. For instance, the impact of colonialism and its aftermath had a tremendous impact on population movements in Tanzania and beyond its immediate borders (Egero, 1974b). One study reveals how the creation of National Parks caused the population to "pile up" (Kurji, 1976); yet other studies (Ishumi, 1984) indicate how the education policy and the quality of rural development in the late 1960's and 1970's had a profound impact on the movement to urban areas.

The initial rapid growth of Dar es Salaam as a primate city was a result of nationhood (Mascarenhas, 1967). Even after the Arusha Declaration, and in the process of following the new economic direction, with the state taking the driving seat in economic management and nearly all the parastatals having their headquarters in Dar es Salaam, the rapid growth of the city during the socialistic era was inevitable. Dar es Salaam now has a momentum of its own to grow for a sustained period. Presently, the process of globalisation, *via* the Dar es Salaam conduit, will increase the primate role of the city even more. For instance, corporations especially in the financial and industrial sectors are more interested in locating their investments in Dar es Salaam as evidenced by the cluster of banks and bureaus de change operating within the city. The major foreign investors are really not interested in following poor people in the rural areas.

It should be evident that confining studies of migration and urbanisation only to the demographic or economic dimensions gives us only a restricted view of the subject. While many demographic studies exist there is a paucity of the latter dimensions. For Tanzania there are only a few specific studies that reflect the impact of policies on migration and urbanization and employment. The national urban study on labour and migration (NUMEIST survey), undertaken by the Economic Research Bureau was a comprehensive survey (Sabot, 1979). There were surveys on employment and earnings carried out in the 1960's. In the 1980's, other surveys were undertaken under the ILO and give a sketch of migration and levels of incomes.[4]

A few individuals have started to take the challenge and move into new areas. For instance, the attitude of people and the situation of land ownership is a matter of great apprehension to many poor people and is bound to have major ramifications (Shivji, 1996). Presently, economic and political changes are having an even greater impact on the way in which the cities will be governed. The importance of urban governance cannot be underestimated; neither should it be forgotten that there are several reasons why governance is so poorly developed (Kironde, 1994). The importance of governance should be recognized with a special section dedicated and added to this volume. While the general answer as to why people move to urban areas needs to be researched, the more specific situation is slightly better.

So why do people migrate? For many general and specific reasons, no effort will be made at this stage to put the movement of people in Tanzania in a global perspective. Instead more emphasis has been put at local level efforts. Table 4.3 systematises this information for the 1980s. A recent study of Babati Town reveals that more than half of the people interviewed had moved into the town because of better employment opportunities.

Table 4.3: *Main reason for current migration*

	Reason for Migration			Returned migrants who had sought work			Probability of Return (if seeking Work)
Destination	To seek work	Marriage	Shortage of land		Total	Other	
D'Salaam	44	15	0	13	72	11	0.20
Other Major Towns	32	7	0	15	54	14	0.30
Local Town	37	7	3	19	66	32	0.46
Agriculture Estate	3	0	2	0	5	6	0.67
Another Village	3	53	18	32	106	13	0.81
Total	119	82	23	79	303	76	0.39

Source: Population Census, 1988

Other reasons for the movement were greater opportunities for survival including coping strategies, marriage and possibilities for agriculture

(Muhandi, 1995). In Mbeya, it was found that migration, natural increase and boundary adjustments were respectively responsible for 46%, 31%, and 23% of the town's population growth. Also, there were more female than male migrants (Mwageni, 1991).

ANALYTICAL FRAMEWORK OF MIGRATION AND URBANISATION

Based on numerous studies in social sciences, an analytical framework of migration-urbanisation revolves around fundamental questions and their variations - How many move? Who moves? What is the origin and destination? What are the "push" and "pull" factors responsible for migration? What is the impact on the areas of origin and destination? It should also be mentioned here that conventionally migration is classified on the basis of whether it is nationally contained or whether it transnational or international. In most countries internal migration is a far more important factor in population redistribution than is international migration. Yet, for several reasons,(including that it is more difficult to study) not enough attention has been paid to internal migration. The big question is: What expectations do people have for migrating? Table 4.4 indicates some of the characteristics and reasons for migrating during the three phases which the country has gone through since independence.

Demography, or the science of population studies has generally tended to shy away from migration because compared to the clear-cut concept of births and deaths and their statistical manipulation, migration offers a much greater challenge. Demographers have been constrained by the fact that "migration" unlike "fertility or mortality" is not an absolute phenomenon but is a transitional one. This challenge has been taken up by other social scientists: such as economists, geographers, sociologists, regional and town planners and even political scientists, all of whom over time have developed concepts, tools, methodologies and models. The concept of studying migration in a holistic way is thus derived from many disciplines.

Relevance of Migration Concepts

The literature pertaining to migration is the least studied, almost everywhere. This is partly because information is more difficult to get, and analysis is dependent on many factors including whether questions were solicited during national census or whether the data was even processed at all. The situation has worsened in recent periods. Thus in the 1967 census, a number of studies emerged about migration and urbanisation from the standard census tabulation (Claeson and

Table 4.4: *Characteristics and expectations of three migration periods*

COLONIAL	POST INDEPENDENCE	POST-STRUCTURAL ADJUSTMENT
Numbers Smaller Number	**Numbers** Increasing Number	**Numbers** Very large numbers
Age & Composition Younger Adults Mostly males	**Age & Composition** Adolescents Some females	**Age & Composition** More Youngsters & Children; in some cases more females
Education Most not been to school	**Education** Most have been to school: Mostly primary Some secondary	**Education** Many not been to school; but also some secondary and even graduates
Targeted goals Cash, Employment	**Targeted goals** Formal sector especially Government & Parastatals	**Targeted goals** Informal sector, casual labour, services and business, movement to non-farm activities.
Period Seasonal & cyclical	**Period** A-seasonal	**Period** Perennial and in groups
Status Migrants had an anchored stake in rural life pertaining to rights for:land, food, social status.	**Status** Erosion of stake in rural life: Villagization removed last vestiges of "tribal security"	**Status** Marginalization of majority' Survival strategies in urban environment.
Freedom & Security Freedoms limited but security through traditional systems.	**Freedom & Security** Many constraints Increasingly through ruling party	**Freedom & Security** Security now mainly through capital and new forms of organization.

Source: Compiled by the author.

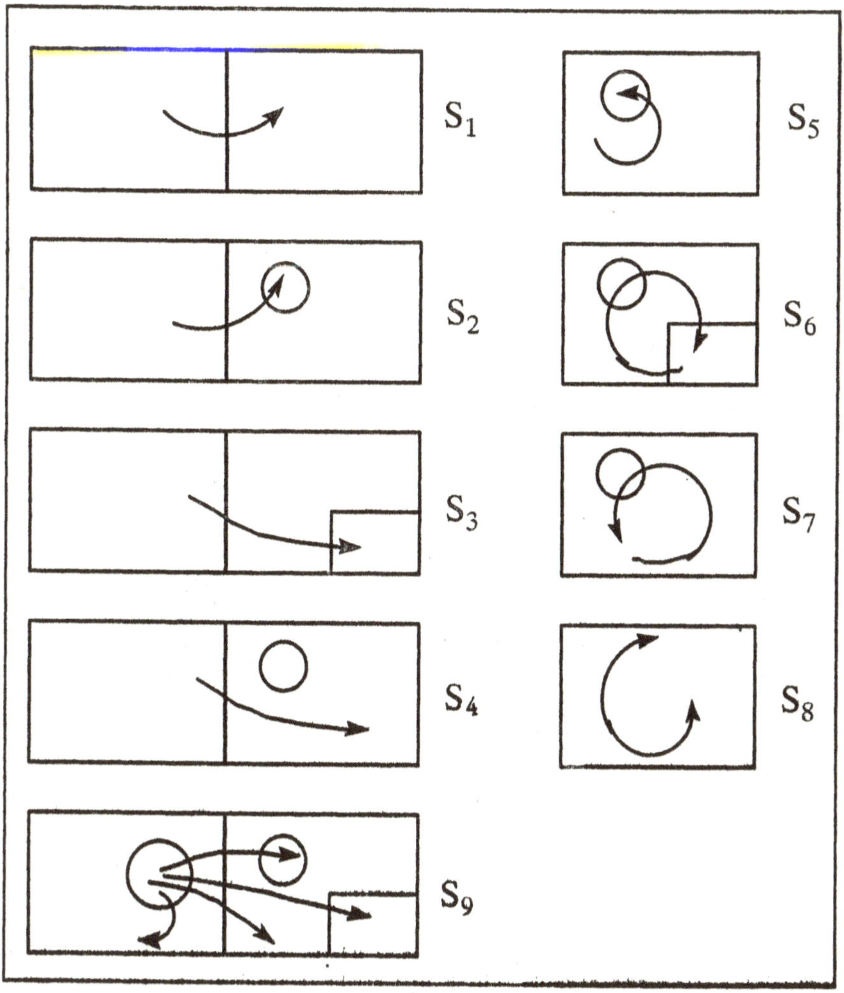

Figure 4.1 *Migration streams in Tanzania (Source: Claeson*

Key:

S_1 - Migration from a region to another region;
S_2 - Migration from a region to the urban centre of another region;
S_3 - Migration from a region to a district in another region;
S_4 - Migration from a region to the rural part of another region(aggregate of S_3)
S_5 - Migration from the rural to the urban area within the same region;
S_6 - Migration from the rural and urban areas of a region to a district in the same region (inluding movement within the district)
S_7 - Migration within a region to the rural areas (aggregate of S_6);
S_8 - Migration within a region(S_5 + S_7);
S_9 - Migration from Dar es Salaam.

Egero, 1973; Mascarenhas, 1973), and additional studies were commissioned (Claeson and Egero, 1972, 1973); there was a sharp decline in the attention given to migration studies in the 1978 census and analysed information from the 1988 census is almost non existent. Few in Tanzania actually appreciate that the country is a major laboratory for the study of migration. A seen from Figure 4.1, several types of migration patterns have been discerned.

TOWARDS A SUBJECTIVE RESEARCH AGENDA

Given the exploratory and sketchy nature of this chapter, this section will not try to identify and prioritize future policy issues, development efforts and research topics, because this would be premature. Since, it is evident from the preceding sections that migration-urbanisation is a very important but complex topic which needs to be addressed immediately, a few topics will be flashed out. These topics are very dependent directly and indirectly on macro-policies of the state and also on micro-factors, which prompt individuals to migrate to towns and cities.

Migration in Tanzania is currently at a new cross roads, in which the sheer magnitude of people involved and the radical transformation of conditions impacting on people demand new and innovative solutions. From the few studies that we have, some strategic research can take place and will yield great returns. For instance the Mascarenhas (1996) study in the two unplanned areas of Dar es Salaam in fact represent models of urban development in which the impetus for growth and investments came from the informal sector. This type of development - informal, rapid, responding to needs, without too much government presence and "planning by doing" is replicated in the entire hierarchy of urban areas in Tanzania. The agenda for migration and urban research is substantial.

The main theory in free migration is differential opportunities, which either pull or push people. The differentials between urban and rural areas are large. A few examples will make the position more clear. The stranglehold of primate cities, in terms of employment, business opportunities and the capacity to absorb funds, is one example. The services that administrative headquarters can offer are another example of the pull factor. For instance, doctors, teachers and other professionals would prefer to be in Dar es Salaam or the other larger areas rather than in the rural areas of Sumbawanga or Lindi. The opportunities in these far away towns are limited, as are services such as schooling, health, etc.

Poverty in Urban Areas

It is only people with some means who can migrate to Dar es Salaam. An increasing number of those who come to Dar es Salaam, unless they find a 'niche,' will not be able to sustain their presence without being a drain on friends and relatives. Most of those who come to Dar es Salaam and the other big towns will not find formal "employment," but will be in the informal sector. New forms of "ties" are emerging in the large urban areas - such as street children, '*machinga*,' street traders, etc.

In contrast, the cost of migrating to a district headquarters or even within the same region is substantially lower. There is evidence of coping mechanisms and safety nets in rural areas to at least partially confront poverty through "primordial ties," be it the extended family, the tribe, relations, clans, place of origin etc, which entitles "kins" to both natural resources and social capital. Many exhibit strong links with the rural areas but look for non-farm opportunities which may become the major sources of livelihood. All these changes are taking place informally because neither the state nor the private sector has really risen to this challenge.

Very little is known about how people cope with income fluctuations and poverty in urban areas. Generally, the cost of food, housing and even some services is higher in Dar es Salaam than in other municipalities and towns. Clearly the advantages and obstacles in urban and rural areas tend to differ, but there are only a few such studies. There are probably over 1,500,000 women in Dar es Salaam, yet so little is known about how they become part of the urban picture (Mascarenhas, 1996).

Many of the indicators of poverty, such as monetary incomes, housing standard, education levels, basic services such as water and health facilities were in the past skewed in the direction of the urban areas. The urban areas were generally regarded as better off. Superficially, the per capita incomes in the urban areas are substantially higher than in the rural areas. In 1993 per capita expenditures in the urban areas were more than 58% higher than in rural areas. Around this period, the per capita income of Dar es Salaam residents adjusted to adult equivalents was T.shs. 158,695. At the bottom of the list were Shinyanga and Lindi with a per capita income of T.shs.14,191 and T.shs.30,243 respectively.

In the mid-1990s, under structural adjustment there were significant changes in the conditions of migration and urbanisation. For instance, there is major deterioration in the quality of life in the urban areas, the gap between rural and urban areas and the intra and inter income differentials are growing substantially. Remedial steps have to be taken because poverty, especially urban poverty, is extremely pervasive.

The amelioration of poverty in urban areas is in an extremely unfavourable situation under the present economic reforms. With so much emphasis given to market forces there is very little explanation as to how these reforms will alleviate poverty and it is unlikely that the private sector on its own will eliminate poverty. Regular employment has fallen drastically. Under the present conditionalities, poverty in the urban areas is bound to increase for three good reasons: first, the preponderance of those who migrate to urban areas are the poor from the rural areas; secondly, in an era of cost sharing, even if migrants could accrue surpluses these would soon be absorbed by the cost of basic needs; thirdly, the inability of the private sector to absorb even a small share of the migrants. Along with this, there is the ambivalence on the part of the government towards the informal sector. This means that for the most part subsistence living has spread to urban areas. Unfortunately this subsistence living carries a very large risk; for the majority it is an attempt to survive without land or means of production.

Urban Governance and Poverty Elimination

With all the political and economic changes that have taken place in Tanzania since the mid-1980s, there is need for major adjustments in the governance of urban areas. Current amorphous policies dealing with urbanisation and migration in Tanzania, are antipathetic to issues of equity and well being, disregard spatial considerations, are oblivious to social engineering and despite all the rhetoric, there is minimal participation in the development of urban areas. Since Dar es Salaam is the main conduit to globalisation and the state has accepted the "virtues of the market" there has been little discussion bordering almost to indifference on the status of rural areas.

There is a big difference between having a city government and having urban governance (McCarney, 1996).[5] Frequently they are used interchangeably. As an evolving term, it means more than having a government that is transparent and accountable. Briefly governance is the "relationship between civil society and the state, between the rulers and the ruled, the government and the governed."

It is not enough to know the national state structures and practices but also the nature of the local government, the opportunities and method for participation, not to mention the role of NGOs and civil society. Civic groups could be related to the private sector, religious and social groups, etc. Governance is also the responsiveness of the city government to local problems. Since, governance is about empowerment it is important to know how they can administer, manage and ensure sound land use and orderly development of urban spatial systems to operate and deliver services under safe and healthy conditions (McCarney, 1996). It enables those who built, constructed, invested and offered needed services to have input.

Governance in the context of poverty also means upholding the rights of the poor to justice and legal protection. For this to happen there is need for additional institutions and people's organizations as well as a balance of action between national, city and the community level. For instance under the old 10-cell arrangement, many misunderstandings and differences could be settled outside the police station or the court of law. At present there is recourse only to the formal judicial system, thus clogging them and making them less efficient.

There has been little institutional capacity to learn from past mistakes and frequently the draconian steps taken in the urban areas, tend to attack the manifestations of the problems and not their root causes. There has been counter tendencies to centralize instead of decentralising, to be top down instead of consolidation from the bottom and local government reforms seem to be in the doldrums. The City of Dar es Salaam exemplifies the problems in its extremes (see Box 4.1).

BOX 4.1: THE CITY OF DAR ES SALAAM

In Dar es Salaam, an elected city government - however inept and corrupt it may have been - was replaced by a central- government nominated Commission in 1996. In the three years of the Commission's existence, its mode of operations, even if they could be justified because of past malpractices, was more noted for its outright bureaucracy, straight jacket decisions and high-handedness. Now the City government more than anytime in the history of Dar es Salaam is beginning to get the capacity to absorb large sums of money, foreign aid, computers, vehicles and consultants. However, the basic problems of the past remain: horrendous difficulties in getting plots, diminishing prospects for the informal sector, little or no capacity for environmental sanitation and management of open spaces and above all growing social and economic problems for the majority of people who happen to be poor. There are glaring inconsistencies in some of the recent developments. How can buildings of 10 and 15 storeys be encouraged when the equipment of the fire brigade finds it a strain to even tackle problems on the fourth floor? Would the coordination not be better if the city did not have to rely on the Ministry of Home Affairs? Does it make sense to privatise every metre of the sea front and deny thousands of its citizens a public good? Is it offensive to have gardens and parks? Having an opposition member in a City Council under a Mayor revealed even to the very hardened that there were serious problems, which had to be tackled. The City Council was abolished and the Commission was installed for an initial six months. It is now approaching three years without any hint of reinstating the Council? Yet there are more roads constructed now than ever before, the city is cleaner, collection of taxes for investment is at an all time high etc. The big question is how can one introduce governance and get development?

This urban bias and the primacy of Dar es Salaam have been a matter of concern since independence. Explicit counter measures were taken to accelerate rural development at first through the transformation approach of capitalised settlement schemes. The Arusha Declaration of 1967 inaugurated an era of major policy interventions. Its aftermath brought sweeping policy changes ostensibly geared towards rural development and control of the economy. Such policies were pursued during the rest of the tenure of the first regime. Landmarks were the near nationalisation of nearly all the major means of production and the financial institutions, the "decentralization" of the administration to the regions and the massive resettlement through villagisation.

There were several direct policies designed to curb the primacy of Dar es Salaam but which had little impact. Three will be mentioned: first, during the Second Five Year it was decided that Dar es Salaam should not get all the major industries but these should be distributed to seven other towns; secondly, in 1974 Dodoma was declared as the national capital and an embargo was put on the construction of any major public buildings in Dar es Salaam; thirdly, to ensure that only bona fide workers and not thieves and loiterers were resident in Dar es Salaam, the infamous '*Nguvu Kazi*' directive was issued in 1983. Most of these policies seemingly only had a fleeting and a marginal impact. However, we need to know what is happening in the other towns.

Large and Small Towns

The primacy of Dar es Salaam at a national level repeats itself, with decreasing intensity, almost entirely in all the administrative levels. Thus, within the administrative structure of Tanzania, all regional headquarters are primate towns or municipalities, in their respective regions, this is also true for the districts. Most services, including critical ones like health, banks, police etc; generally are only available in-towns. The importance of understanding the role of small towns in this part of the world has been on the agenda for some time (Simon, 1992).

Good Neighbour Towns

Because we have so many neighbours, attention has to be paid to border towns. Such research has to be taken in the following three perspectives: (a) growing emphasis on regional integration especially of economically and culturally linked areas; (b) growing demand for border region cooperation because of increasing competition, conflicts and management of common resources; (c) the rights of border people to develop. Should the peripheral character of border areas condemn them forever?

These towns can spearhead commerce, trade, services and agriculture and resource use. What services do they presently have? The government's role should be to create an enabling environment so as to encourage local initiatives. Presently, to trade formally has so many bottlenecks that it encourages smuggling, profiteering, petty trading etc. Likely candidates could be Tunduma, Namanga, Kigoma, Isaka, and the larger sea and lake ports.

Managing the Economy and Natural Resources

Prior to the present changes instituted through structural adjustment programmes, a new form of a dual economy had functioned in Tanzania for at least two decades under very different circumstances. The "parallel economy," the "second economy,"[6] "the informal economy" "survival economy" are all terms that acknowledge this reality. The national or formal economy is the one that has drawn most of the energies both of the government, the donors etc. Irrespective of whether the Government has to pay its debts, allow market forces to dictate terms of trade, transfer of technology or agree to cut down on social services, the other economy is of direct relevance to the vast majority of people in Tanzania. This is simply because people have basic needs and a basic determination to get the best out of life.

It is in the city of Dar es Salaam that the sharpest confrontation, compromises and absurdities take place. For instance, billions of tax payers shillings are "used" to construct towers and slabs, so that the shiny vehicles of those in power can be sheltered for a few hours! Yet, there are no funds for desks, clean water and basic drugs. The largest number of children, even in the urban areas, die because of preventable water borne diseases. The anomaly of Dar es Salaam is well known. Although, there are links between the formal and informal economy these are still grey areas.

It would appear that a similar process is taking place on an even bigger scale concerning food and the use of natural resources. Thus while within the city, very little food is grown, hundreds of tons of grains move across districts and regional frontiers, natural resources of all kinds are used and services are provided for millions. We know very little about this rural-urban economy, not even the volume and the value of trade at the Tandale Market in Dar es Salaam. Yet, what takes place in Tandale is replicated in a small way from Ifakara to Kabuku, Gairo to Mbarali.

While the major international investors are being wooed, with limited success, there are legions of small investors and entrepreneurs, who have oiled the economies of the rural areas and those of the small towns. One has only to trace ownership of stores and shops in Shinyanga, Kahama or Singida to appreciate the change that is taking place. Thousands of individuals have entered into the local economy, exploited new opportunities, taken risks, and thereby provided much needed employment. Some have expanded, others diversified their businesses while some have even collapsed. These thousands of individuals can best function in small towns.

There is another dimension that we need to explore. As Tanzania tries to make its economy grow, several of its resources are being exploited: minerals, natural resources, tourism. At another level the development of the infrastructure, roads, ports, hydropower, to mention only a few, all lead to the creation of strategic centres. There is need for a topology of these centres and attention has to be paid to their role, function and the services they provide or could provide. A set of examples of small town development is abridged in Box 4.3.

Urban Settlements and Resource Management

This section focuses on natural resource management and urbanisation. In areas of diminishing resource availability, or adjacent to high value resources, or in areas with high population density clusters, the presence of even a small town can make all the difference. It has been observed in Lushoto that centres like: Bumbuli, Soni, Mlalo, Mtae and others actually become small towns which encourages trade, communication, diffusion of information etc. To some extent, such townships reduce pressure on natural resources. In addition, they expose people to products and to new opportunities which otherwise would not be known. An area worth researching is the link between traditional organisations and management and how they respond to present day needs. How can the services they provide be up-graded?

Urban Settlements and the Tourism Sector

Tanzania is poised to be a tourist destination country. At the moment it is most ill prepared to realize its potential. It is ironical that some of the poorest people in Tanzania live in or around some of the most unique places (Ngorongoro, Rufiji) in the world. This also applies to historical towns along the coast which could attract cultural tourists. Aping Kenya is an easy option.

Who should be involved in planning for the development of these towns?[7] How could the multiplier effects be more focused to bring change? What can it do for employment and cultural development? Can some of the stagnation in some of the old towns be turned around? Candidate urban areas include Bagamoyo, Pangani, Karatu, Mto wa Mbu, Kilwa etc.

BOX 4.2: A TALE OF TWO TOWNS: BABATI & MAKAMBAKO

BABATI is a thriving agricultural town which has grown rapidly from 9759 inhabitants in 1978 to 21794 in 1988. It is astride the Great North Road and is also at the cross- road for traffic going to the west, and to Rwanda, Democratic Republic of Congo and Burundi.

In a survey of 255 households it was found that over 55% had migrated into Babati mostly from within Arusha Region but a large number coming from the relatively poor Dodoma and Singida Regions. Over 76% and 11% had primary and secondary schooling respectively. Nearly a third were drawn by the agricultural potential and indeed 287 individuals or 24.2% of the sample claimed to be farmers. The majority came to look for employment and nearly two thirds are formally employed in the private sector or the government.

What do the inhabitants do? 50% of the households are farmers, running shops (12.5%) and about 10% each in petty trading, and technician, administration and professionals services. Not surprisingly because of its resources richness a great variety of crops do well including wheat and maize, pigeon peas, beans, sugar cane, sunflower, cotton, coffee and a variety of horticultural crops. The very large NAFCO holdings, which produce wheat on a mechanised basis, are also nearby. As a bonus, the 350 to 430 fishermen in Lake Babati can in good years, produce over a thousand tons of fish.

Nearly two thirds of the migrants have been in Babati for over 15 years, another 26% arrived 5-14 years ago. Most want to reside permanently but still have links with their previous homes. That they are in a town and not the countryside can be gauged by the numbers working in: Retailing (271), stall holders (48) restaurants (36), guest houses (34), bars (30), local brewing (20), butcheries (20) but the largest occupation is schooling (368).

Source: Muhandi (1995)

MAKAMBAKO, is a transport node. Within a very short time its population grew from a few households, to 9,097 in 1978 to 15,542 individuals in 1988. It started as a construction camp for workers of the Railway to Zambia (TAZARA), the Highway (TANZAM) and the pipeline (TAZAMA) as well as a major road going south to Njombe and another going to Mgololo, site of a paper mill and one of the country's major industrial investments.

The town grew spontaneously and by the 1990's had the following:250 shops, 56 restaurants 7 guesthouses, 4 petrol stations, 37 bars, several butcheries and many stall holders in the market place and elsewhere. Apart from the market and service activities, there is an evolving process of locally sponsored industrialisation such as furniture making, construction, repair and processing activities.

Formal employment is very constrained and this means that the majority of migrants must rely on the informal sector: petty trading, brewing of local beer and wine. The majority came to do business or waged labour (24%) but almost as many came to stay with friends/ family. Over 11% came to farm. Links with the rural areas are not severed but maintained through parents or children being left behind and through frequent visits. Most of the people are urban dwellers with a rural lifestyle: growing their own food (95%) having more than one job, including farming (96%).

Although Makambako Town is a valuable link between the rich and productive rural areas in the south and the region and nation, there have been little effort by the government to invest in the basic infrastructure including in plans and layout of streets, plots, water, electricity, and sanitation. Even with reforms and trade liberalisation, a town minus the infrastructure, without credit facilities, no encouragement to the private sector cannot optimally serve its hinterland or its people.

Source: Holm (1992)

Towns, Mines and Management of Natural Resources

Several towns in Tanzania owe their origins to mines: Chunya for gold, Mwadui for diamonds, Uvinza for salt. The mining sector is Tanzania's major growth sector. With a bit of forward planning several growth areas can be identified, planned for and made to function as areas of rural development (see Box 4.3). Presently, nearly all areas of artisanal mining, be it gold mining or mining for Tanzanite, rubies and other precious stones, can develop into large slums. Not only are they an environmental and a health hazard but socially such centres can be very insecure and unpleasant for those residing in them.

BOX 4.3: TOWNS, MINES & MNR*

In its heydays Mwadui was essentially a mining town replete with infrastructure: a health centre, electricity, running water, even an airfield. Hundreds of local people worked in the mine. At its height, some $15 million dollars worth of diamonds were being exported.

What are the plans for the centres which will produce tens of millions of dollars worth of exports of gold, precious stones and minerals? How can the opportunities emerging be translated into benign and forward looking actions rather than a repetition of evictions and exploitation, more reminiscent of the Wild West.

It is necessary to look afresh, at policies on management and use of natural resources, fair compensation for people and lasting social investments in the areas of mining and resource exploitation. It should not be forgotten that such centres provide rural areas with new opportunities for agriculture, employment etc.

* Management of Natural Resources

Establishing an Urban Hierarchy for Tanzania

The two capitals of the nation, Dar es Salaam and Dodoma, and the capitals of the regions and districts are all relatively well known. Less well known are the centres which provide services for local communities and others, even less known are villages, some of which have very large population cluster s(Ikwiriri has more than 20,000 inhabitants). During the 1967 it was found that while officially there were only 32 "towns," but in actual fact there were 204 centres that offered services (Mascarenhas, 1973). Is it true that there are over 435 trading centres in Tanzania? If this is true, within the 883,621 km^2 of the country, each centre roughly serves 2031 k.m^2; and are the centres dense enough to lead to development? What is the pattern in the country?[8] Do centres in the dry areas tend to be far apart? What are the consequences?

Towards Understanding the Urban System

The role, function and implication of small and medium sized towns is becoming a critical necessity. Since the circumstances of the recent past are no longer seriously relevant today, the reluctance on the part of the government to establish towns and to pay attention to physical planning cannot be justified any longer.[9] Since gathering information on urbanisation and migration is expensive, with the required political will, such data could be collected when the postponed census takes place. Meanwhile, since a functional approach should be used to classify the large, median and small sized towns, there is urgent need to carry out strategic research before the next census. For instance, are divisional headquarters of productive areas more likely to grow into small towns? Or is it entirely a function of transport, markets, population size etc? What has been the influence of the private sector *vis-a-vis* the public sector?

Appropriation

All over Tanzania, the appropriation of resources is taking a dramatic turn. It is often assumed that the differentials are greatest in the large towns. It is worth researching in this area. The exclusion of people from their traditional rights deserves attention (see Box 4.4).

BOX 4.4: MISAPPROPRIATION AT A LOWER LEVEL.

It is assumed that social and economic differentiation is more acute in the large urban areas and less so in the smaller centres. This assumption is not entirely true. At Giting village, five households have managed to appropriate most of the land. Most of the communal resources including land for public use, the silo, the truck and even the piped water scheme were misappropriated. Social stratification is taking place, use of resources both natural and human are not reinvested or redistributed and this retards local development. By linking with the nearby village of Dirma and with district and regional capitals, and by corrupt practices they have a stranglehold on the economy of the area.

Source: Loiske (1995)

Gender Aspects of Migration and Urbanisation

In recent years several studies have emerged on the gender aspects of migration and urbanisation in West Africa (Baker & Aina, 1995). However, there is a great dearth of research on the gender aspects of migration and urbanisation in East Africa. Most of the researches tend to be once for all studies. If one wants to study changes under unstable

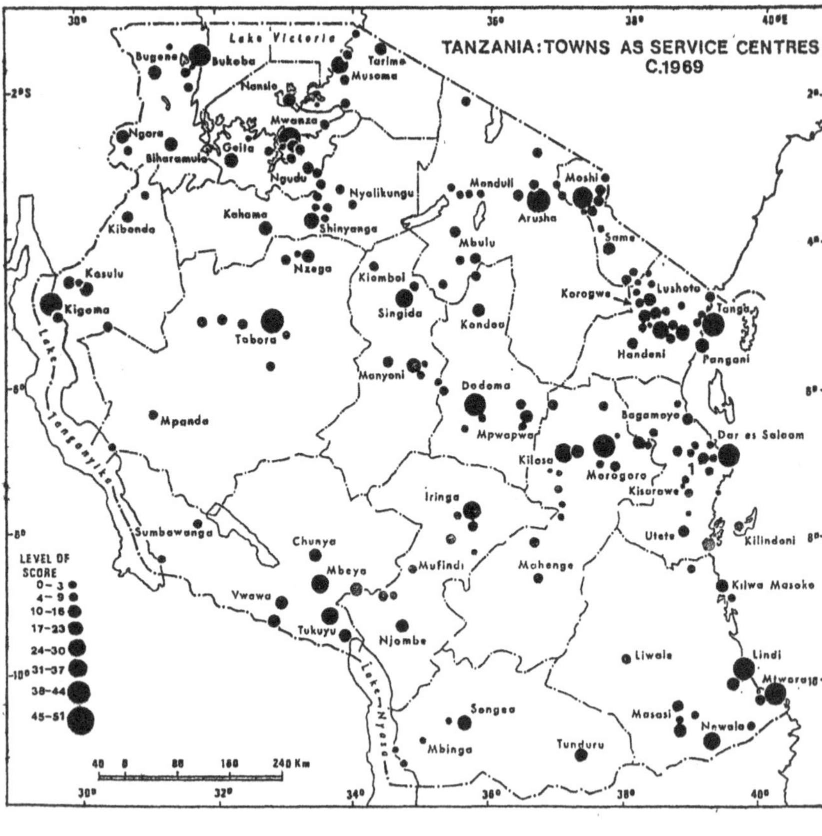

Figure 4.2: Service centres in Tanzania

situations and the coping strategies of different groups, it is worth carrying out longitudinal studies. There is one good baseline study on how to do this in Dar es Salaam.[10] Another aspect of gender studies would be to evaluate the increasing number of young female migrants. A few specific demographic studies are now available, for instance on barmaids (Lihawa, 1994); commercial sex workers (Kamazima, 1995); and female house-helpers (Mtingwa, 1996).

CONCLUSION

This chapter on migration and urbanisation in Tanzania has tried to capture some aspects of the situation in Tanzania. The main sources of information on migration and urbanisation have been reviewed. It was noted that most of the information on migration is derived from censuses and almost none on surveys. This means that in a situation of rapid changes, there are frustrating gaps and synthesis is made difficult. On the whole demographic aspects have been relatively well covered compared to most of the other fields.

Urbanisation and migration are not only a matter of demography. Therefore, an attempt has been made to indicate some of the processes at work. Within the context of the socio-economic processes an analytical framework was sketched. Migration and urbanisation are taking place because of economic and political differentials and the drive by individuals to have a better life. The impact of internal and external pressures manifests itself differently. Migration and urbanization are influenced by development and environment and in turn have an impact on them.

The fifth part, which covers a large part of the chapter, is a a twelve point subjective agenda for better understanding the important role of migration and development. The agenda is large, but at a time of great changes it requires that research, policies and styles of management be altered to suit the changing circumstance of handling opportunities for development. It demands participation of the civic society, tolerance and the sharing of power, a sense of history and an appreciation of the value of plurality and justice. Attention should be focused on "institutional development, housing and infrastructure, grassroots financing, health, education are all needed" (Tannerfeldt, 1995). In this respect social assistance could well be the new sector for integrated rural and urban development.

END NOTES

1 In the creation of a sovereign, republican state, local governments authorities, with chiefs, a large alien minority concentrated in urban areas and multiple parties were considered as an unnecessary distraction. Therefore "chiefs" were abolished first, local governments and town councils followed. To curb the economic power of the landlords, buildings were nationalized, the "decentralization" or really centralization of the powers of the state and the single party was established. Only then did the government get involved in the villagization exercise of the mid-1970's.

2 New evidence has been unearthed that show that small towns go even further back than the major Arab incursion. The new evidence of large settlements go back to the Greco-Roman period (Chami, F. Department of History, Archaelogy Unit, University of Dar es Salaam ; personal communication. 1997)

3 The six components are: (a) administrative experience; (b) sustainability; (c) balance; (d) self reliance, (e) equity and (f) empowerment (Bryant & White 1982). On the other hand Todaro defines development in terms of: (a) raising people's living conditions; (b) growth of people's self-esteem and;(c) increasing people's choice (Todaro 1985).

4 An important observation that can be made is that the major assets of rural Tanzanians are the three "Ls"; land, labour and livestock. Land cannot be taken to urban areas and to that extent is not a disposable asset. Both labour and livestock can be sold for cash.

5 See McCarney P. L. (1996) "Considerations on the Notions of 'Governance' - New Directions for Cities in the Developing World," in in Mccarney, *Cities and Governance*, University of Toronto Press, Toronto, p. 208

6 *The Second Economy in Tanzania* by Malyamkono, T.L'& Bagachwa M S D, is the best known and still the most comprehensive source available.

7 Through the Friends of the Museum, a panel discussion was held in July 1998. Participants who attended included historians, civil servants, local investors, students, donors, contractors, architects, students. The meeting also provided a forum to explore areas of common interest.

8 During its 1969-74 Development Plan, Kenya had some 1,400 Rural Trade and Production Centres. It was argued that the district capitals were adequate for a colonial class with large land holdings but inadequate for the needs of small-holder farms.

9 There were nine district councils which were in the process of being established: Korogwe, Mpwapwa, Kilosa, Njombe, Kondoa, Kahama, Tukuyu, Bagamayo, Muheza. There were a total of 19 urban councils, i.e 1 city; 9 municipalities; 9 town councils. (Semboja & Therkildsen, 1991).

10 Two unplanned settlements, out of the 44 in Dar es Salaam were studied comprehensively with a gender bias. There was very little government assistance and most of the activities were informal. The two communities had a joint population of nearly 19,000 people,approximately the population of a largish district town. The striking feature is the level of informal activities which enable the people to survive. Most of the women see many advantages in the areas they live compared to the rural areas. Over 70% were migrants. (Mascarenhas, O. 1996).

BIBLIOGRAPHY

Adepoju, A. (1995)
"Migration in Africa - An Overview," in, Baker, J. and Aina, T.A., *The Migration Experience in Africa.* Nordic African Institute, Uppsala, p. 87-108.

Aeroe, A, (1992)

Amin, S. (1995)
"Migration in Contemporary Africa: A Retrospective View" in Baker, J.and Aina, T.A. (eds), *The Migration Experience in Africa,* Nordic African Institute, Uppsala, p. 29-40

Baker, J. and Pedersen, P. O. (eds), (1992)
The Rural - Urban Interface in Africa. Sem. Proceedings No 27, Scandinavian Institute of African Studies, Uppsala.

Baker, J and Aina, T. A. (eds.), (1995)
The Migration Experience in Africa. Nordic African Institute, Uppsala, Sweden.

Campbell, J. and Pons, V. (1987)
"Urbanisation, Urban Planning and Urban Life in Tanzania" *Occassional Paper* No. 4, University of Hull.

Claeson, C. andEgero, B. (1971)
"Movements to Towns in Tanzania: Tables and Comments." *BRALUP Research Notes* No. 11.1; BRALUP, University of Dar es Salaam, pp. 91.

Claeson, C and Egero, B. (1972)
"Migration and the Urban Population: A Demographic Analysis of Population Census Data for Tanzania" *BRALUP Research Notes* No. 11.2; BRALUP, University of Dar Es Salaam.

Claeson, C. and Egero, B. (1973)
"Migration" in Egero and Henin, (eds.), *The Population of Tanzania,* BRALUP and The Bureau of Statistics, Dar es Salaam.

Demographic Unit, (1997)
Abstracts of *M.A.* (Demography) Dissertations: 1987 - 1996, Demographic Training Unit, University of Dar es Salaam, Dar es Salaam.

Doherty J. M. (1976)
"Ideology and Town Planning in Tanzania." *Journal of the Geographic Association of Tanzania,* No 14, June, Geography Department, UDSM, Dar es Salaam, p. 79-104.

Egero, B (1974b)
"Population Movement and the Colonial Economy of Tanzania" *BRALUP Research Paper* No 35; BRALUP, University of Dar Es Salaam.

Gould, W. T. S. (1995)
"Migration and Recent Economic and Environmental Change in East Africa," in, Baker, J. and Aina, T.A. (eds.), *The Migration Experience in Africa.* Scandinavian Institute of African Studies, Uppsala p. 122-145.

Holm M. (1992)
"Survival Strategies of Migrants to Makambako - An Intermediate Town in Tanzania," in, Baker and Pedersen (eds), *The Rural - Urban Interface in Africa.* Sem. Proceedings No 27, Scandinavian Institute of African Studies, Uppsala.

Ishumi, A. G. M. (1984)
The Urban Jobless in Eastern Africa, Scandinavian Institute of African Studies, Uppsala, Sweden

Kamazima, S. R. (1995)
"Existence and Implications of Female Commercial Sex Workers in Urban Centres: A Case of Female Streetwalkers in the City of Dar es Salaam." *MA* Dissertation, Demographic Unit, UDSM.

Kironde J.M. L. (1994)
"The Governance of Urban Development in East Africa: A Research Perspective." Workshop on Urban Governance, Dar es Salaam.

Krokfors, C. (1995)
"Poverty, Environmental Stress and Culture as Factors in African Migration." in Baker, J. and Aina, T. A. (eds), *The Migration Experience in Africa.* Scandnavian Institute of African Studies, Uppsala, p. 55-63.

Kurji, F. (1976)
"Conservation Areas and Their Demographic Setting in Tanzania." *BRALUP Research Report* No 18, (NS), University of Dar es Salaam, Dar es Salaam.

Lihawa, H. A. (1994)
"Rural-Urban Migration in Tanzania: A Case of Barmaids in Dar es Salaam City." *MA* Dissertation, Demographic Unit, University of Dar es Salaam.

Loiske, V. M. (1995)
"Social Differentiation, Conflicts and Rural-Urban Interaction." in Baker & Aina (eds), *The Migration Experience in Africa,* Scandinavian Institute of African Studies, Uppsala, p 221-233.

Mascarenhas, A.C., (1967)
"The Impact of Nationhood on Dar es Salaam," *East African Geographical Review,* Vol 5, Kampala, p. 39-46.

Mascarenhas, A.C., (1973)
"Urban Growth," in, Egero, B. and Henin, F. (eds), *The Population of Tanzania,* BRALUP/Bureau of Statistics, Dar es Salaam, p. 76-97.

Mascarenhas, O. C., (1993)
"Population and Development in Tanzania: A Review of Information Sources," Report Prepared for ILO/UNFPA and Sponsored by the Ministry of Economic Affairs and Planning, DUP, Dar es Salaam.

Mascarenhas O. C. (1996)
"Gender Aspects of Urbanisation and Natural Resource Management in Tanzania." Mazingira Institute, GUE Project, Nairobi.

Mbonile, M.J. (1993)
"Review of Sources of Information on Voluntary and Forced Migration in Tanzania." *Occassional Paper* No. 6, Demographic Unit, University of DSM, Dar es Salaam.

McCarney, P. L (1996)
"Considerations on the Notions of 'Governance' - New Directions for Cities in the Developing World," in, McCarney, P. (eds); *Cities and Governnace,* University of Toronto Press, Toronto.

Mlay, W. (1981)
"Assessment of Inter- and Intra-Regional Migration in Arusha Region." Mimeograph, University of Dar es Salaam, Dar es Salaam.

Mtingwa, M. (1996)
"Socio-economic Effects of Female Migration in Tanzania: A Case Study of House Girls in Dar es Salaam Region," *MA* Dissertation, Demographic Unit, UDSM.

Mwageni, E.A. (1991)
"The Components of Population Growth in Mbeya Municipality, Mbeya Region Tanzania,"
MA Dissertation, Demographic Unit, UDSM.

Rafiq, M. (1988)
"Urbanisation: An On-Going Process," in, *1978 Population Census*, Vol VIII, Bureau of Statistics, Dar es Salaam. p. 181-210.

Rwechungura, C. (1988)
"Commuting as a Factor of Census Under-enumeration in Tanzania: A Case Study of Bukoba Town," *MA* Dissertation, Demographic Unit, UDSM.

Sabot, R.H. (1974)
Economic Development and Urban Migration: Tanzania, 1900-1971, London, Oxford University Press.

Shivji, I.G. (1996)
"Report of the Presidential Commission of Inquiry into Land Matters." Scandinavian Institute of African Studies/MLHUD, Uppsala/Dar es Salaam.

Simon, D. (1992)
"Conceptualising Small Towns in African Development in South East Africa," in, Baker and Pedersen (eds), *The Rural - Urban Interface in Africa*. Sem. Proceedings No 27, Scandinavian Institute of African Studies, Uppsala, p. 29-50.

Tannerfeldt, G. (1995)
"Towards An Urban World." Swedish Agency for International Development, Stockholm.

URT, (1990)
1988 Population Census, National Profile, the Population of Tanzania, the Analytical Report, Bureau of Statistics, Planning Commission, President's Office, Dar es Salaam.

Zelinsky, (1971)
"The Hypothesis of Mobility Transition," *Geographical Review*, Vol 10 p. 451-488.

5

URBAN GOVERNANCE AND URBAN PLANNING IN TANZANIA

*Fred S. Lerise**

INTRODUCTION

Tanzania, like many other African countries, is carrying out reforms towards economic liberalisation and political pluralism, aimed at more open, less exclusive and accountable institutions and practices. Within these reforms, it is not only a strong government that is the most important factor, but also a strong and well organized civil society which can relate positively and effectively with the government and the public sector. This chapter argues that for the reforms to become effective and productive, more attention is needed on the relationship between government institutions, civic groups, the business community and individuals. Both urban governance and urban planning should be organised in such a way that they contribute and enhance the reforms.

Through the governance concept, structures are conceptualized as processes and relationships. By analysing urban planning or the management of urban development using the governance paradigm, we are directed to the relationships between different stakeholders. It is through such interactions that urban areas are created and managed or mismanaged. We shall show that, reforms in planning processes have been rather slow particularly at the central government level, and the relationship between state, civil society and the private sector is still characterised by state control, subordination, exclusion, lack of transparency and accountability.

Since urban planning is both a subject and a tool for the reforms, it is

* Senior Lecturer, University College of Lands and Architectural Studies, UDSM

considered limiting to measure the Tanzanian urban planning enterprise by the national reforms. To avoid measuring planning by itself, we shall therefore evaluate planning against practices and theories in other parts of the world, drawing examples from societies searching for more democracy.

Since its inception, urban planning has been evolving. One aspect of planning which has attracted attention more than others is the question of public interest. Although planning theoreticians are yet to offer acceptable ways of defining and protecting the public interest against individual interests, attempts made so far at global level have made positive contributions towards understanding the problems associated with the handling of public interest better than was the case in the 19th century. The urban planning enterprise at global level has developed techniques and ways through which the public interest can be better identified and protected. Stories about participatory and collaborative planning practices are increasing. It is also reported that such practices have improved urban management processes and contributed positively in the management of human settlements in various parts of the world. How far is the Tanzanian urban planning practice in the reform agenda? How can we explain that position? These are questions dealt with in this chapter. We start by mapping the conceptual landscape of the terms 'governance' and 'urban planning,' after which we shall summarize the main principles for good urban planning in societies aspiring for decentralized, democratic, open and inclusive systems of governance. We shall later apply the principles and use stories to locate the Tanzanian urban planning practice.

MAPPING THE CONCEPTUAL LANDSCAPE

Governance as a Concept

Like many concepts developed from, and for, western societies, governance is slowly finding room in the Tanzanian urban management debate. The literature and especially official reports interpret the concept in different ways. For instance, the National Human Settlement Development Report criticises local government reformers in Tanzania for their narrow understanding of the concept of governance. The Report argues:

> Most of the proposals that have been advocated for improvement in local governance have conceived governance in the sense of "good government." There has been an undue tendency to concentrate on state organs as opposed to the relationship between ruler and ruled, and the various actors that shape local governance. The idea of governance however transcends good government and involves all the various actors (the civil society) on the local scene. Government policy towards recognizing and incorporating the activities of the civil society (e.g. through NGOs) is still weak. One area of improvement could be to extend the definition of governance (URT, 1996: 38).[1]

In the strategic plan for restructuring the Dar es Salaam City Council into three municipalities, the term governance has been extended to mean:

> The totality of how government is done and what it achieves. *Good governance is government* carried out by elected persons regularly called to account by elections... which involves *participation by institutions of civil society in the process of governing* and which aim to maximise the welfare and potential of all the people (URT,1997: 132).[2] (emphasis added)

The above conceptualisation of governance relates closely to the classic liberal democratic model which emphasises the nature of the state, particularly how to put the state institutions in place and how to make them accountable to the citizens through elections. Putting good governance in place and making it accountable is in that conceptualisation, the responsibility of civil society. At a more general level, a wider conceptualisation of urban governance is provided by Hyden (1992) as quoted in Swilling, (1998) as the management of relationships linking different actors in urban development. While emphasising the interrelationships, that conceptualisation of governance pushes the debate closer to the fluid associations across public-private community divides and voluntary associations within civil society.

Following that approach to governance, the actual management of local government services, infrastructure and urban development should be decentralised. According to Halfani (1997),[3] the main idea in the governance paradigm is the recognition that urban development is a result of complex interactions between public institutions, civil societies and the private sector. For instance, the creation and management of urban space, which is one of the concerns of urban planning, is a result of individuals or groups as well as of government policies and actions.

The governance paradigm therefore directs attention to the complex set of vertical and horizontal relations among different actors in the urbanisation process and its management, particularly focusing on the contributions which come, or are likely to come, from hundreds of community organizations as compared to what a single government can contribute by itself. Hyden's conceptualisation of governance fits well with the growing concern among urban planning practitioners and researchers who call for an understanding of urban planning not as a monopoly of government institutions but as a shared responsibility among a multitude of actors including the different levels of government. In that understanding of urban planning, the role of planning has to be widened to include not only developing ideas on what should be done to facilitate urban development, but also how that should be done and within what form of institutional framework. Planning has to contribute towards charting an institutional framework, which provides citizens with the powers to ensure that their interests, needs and rights are respected and taken into account in the process of managing urban development. However, as we shall argue in the following sections, the new scope of urban planning is yet to be fully adopted and practised in Tanzania.

In the following section we discuss the extent to which urban planning in Tanzania has been reformed towards the good governance vision. Before we open up the discussion on the status of urban planning in Tanzania, it is important to summarize the main elements or pillars of good governance within which urban planning is to operate. The pillars are:

- Close engagement between the state, civil society and the private sector in planning and urban development;
- Decision making arena, which is open and accessible to state as well as to non-state actors;
- Institutions and management systems that facilitate consensus building and conflict resolution processes and which are open to different actors.

These pillars of governance require that urban planning be also reformed to not only contribute towards the good governance vision but also operates within that type of governance. How Tanzania conceptualises urban planning within the governance paradigm is discussed in the following section.

URBAN PLANNING IN TANZANIA

In Tanzania, urban councils and other local authorities operate with two types of plans. The first are master plans, structure plans or local plans and more recently strategic plans, which traditionally focus on land use and infrastructure. These plans are not linked to the annual budget but do provide a framework to regulate land use and the overall built-environment of an urban area.

The other types of plans are sector or departmental annual plans, which are linked to the local authority budgets. Such sector plans make up the annual plans for a given council. In this discussion we shall pay more attention to the first type of plans.

By its nature and its concern with equity and public interest, urban planning is a tool for governance. It is one of the ways of managing urban development. In many societies the meaning of, and the way in which urban planning is carried out results from the prevailing understanding of and the type of governance pursued. In societies where governance is understood as good government, urban planning is centralised and monopolised by the central and the local states. On the other hand a society which aspires for good governance organises its structures and decision-making processes as well as its urban planning enterprise in accordance to, and in order to, foster good governance principles. To illustrate this point we use an urban planning practice example from a society which understands governance as the way the state relates with its civil society, and has organized its urban planning practice to take into account societal diversity in its planning decisions. We use an example from Canada.

Qadeer (1994) provides a story of how the Canadian urban planning practice was applied to diffuse a public controversy over housing development in the city of Vancouver. He narrates:

> In the 1980s, Vancouver aggressively elicited capital from Hong Kong and Taiwan. Immigrating entrepreneurs and professionals then fueled the real estate boom, driving up housing prices. In Kerrisdale and Shaughnessy, which are leafy Victorian neighborhoods, Asian immigrants rebuilt homes to suit their tastes and activities. Within the zoning envelop, (regulation) the newcomers used the permissible coverage and setbacks to create 'massive and lavish' homes nicknamed Monster Homes. The long-time residents fiercely demanded that the city prohibit such development. The controversy simmered for about two years, often assuming ethnic overtones.
>
> The planning system had been caught by surprise, since the new homes conformed to the regulations.... The Mayor and the City Council designated the neighborhoods as a special study area and asked the planning department to review the RS-1 Zoning. The Planning Department held public hearings that brought together representatives of the original residents and the new Asian residents to hammer out a compromise resulting in the adoption of new (1993) design guidelines. They required builders of homes to take into account streetscape, the architecture of adjacent homes, and the landscape. The maximum allowable sizes was unchanged, but brought under design control.[4]

Through urban planning initiative, the City Council and the two civic groups had a forum in which their different and conflicting interests were negotiated and consensus arrived at. Planning was used as a tool for resolving housing development problems and possible social conflict through negotiations.

On the assumption that the social, political and economic reforms taking place in Tanzania since 1980s are meant to pursue the goal of good governance, we can then ask, how far has urban planning in Tanzania been reformed and re-organized to produce stories like the Canadian one? A more productive way of dealing with this question, which, in my view, is central to the state of the art in urban planning, is to look for theories of organizing the urban planning enterprise to serve as a tool for good governance. We now turn our attention to the literature on good principles of urban planning.

Urban Planning: Global Level

Western-based literature suggests that urban planning was invented in Europe in the 19th century as a tool to control urban development. It was meant to deal with problems of public health, poor and overcrowded housing conditions, high rates of crime etc., resulting from the industrial revolution. Urban planning was then a way of facilitating government intervention to ensure that building regulations are followed, to guarantee at least minimum standards of living conditions.

At present almost two centuries later, the purpose of urban planning has not changed significantly. Its main goal is still to protect public interest which, according to Devas and Rakodi (1993)[5] is to restrict the rights of private landowners to use their land as they wish. That restriction is carried out for the interest of the society at large and because governments have a social and political responsibility of assisting the poor. Additionally, planning is also used to protect property owners and local business interests against the ravages of unrestrained market rationality - through land use planning, zoning, anti-pollution planning etc.(Friedman, J. 1987) [6]

Although the scope and purpose of planning is always widening, still the central issue or the main area of concern is the public interest. Theoreticians, researchers and practicing planners wishing to improve the performance of their craft, now and then focus their attention and resources searching for answers to the question: how best to define the public interest and ensure that it is achieved and safeguarded. Attempts to address the problem of public interest are well documented in the literature. (see for example, Paris, C. (1982),[7] Healey, P., *et. al.* (1982)[8] Bryson, J. M. and Crosby B. C.(1992)[9] etc). Such attempts to reform urban planning are more pronounced in the Western world where good governance objectives have been pursued for a longer period than the rest of the world.

Most theoreticians have targeted their discourse in changing the way urban planning is carried out. For instance there are arguments to change from the plan-led regulatory planning approach pursued by state institutions to that carried out through communication and collaboration among different actors, as outlined in the Vancouver case.

Famous planning theoreticians like Habermas (1989), Friedmann (1992), Healey (1996), and Forester (1993), have been interested with how to improve urban planning. Though writing from their different cultural, social, economic, political and environmental situations, they seem to be moving towards a similar conceptualization of urban planning. They recognise the failure of the planning approach guided by the scientific rationalist paradigm, because of the assumption that societal diversities can be scientifically and rationally reduced into one position on the basis of which decisions are made.

One way of recognising and taking societal differences into account is to open up that planning arena to different stakeholders and let them participate in defining the public interest and chart out ways to achieve it. Studying urban planning in the Danish town of Aalborg, Flyvbjerg (1998), shows that categorisation of stakeholders into government, business community and civil society and laying down procedures to ensure that the three spheres are represented does not guarantee that the real public interest will be pursued. To illustrate this point, he writes:

> Planning was used by the most powerful stakeholders, in a highly closed decision making process dominated by a small, almost exclusively male, elite of top politicians, high ranking civil servants and business leaders supported by the local press and the police. The division here is not between civil society and the state but between those with power and those without power. The most powerful part of civil society - the local business community represented by its leaders - collaborates with the most powerful politicians and administrators, i.e. the mayor, the alderman and the city managers, in order to shape the city to their own interests.... The rest of civil society, the unorganized citizen and even the City Council have only marginal influence on planning and policy decision and these parties are actively kept on the margin (Flyvbjerg, 1998).[10]

Coalition among the most powerful actors from the three spheres hijacked the planning process and determined the outcome of the plan in their favour. We learn from the story that, it is not sufficient for the reforms in each sphere or some of the spheres do constitute a force, which should be taken into account.

At a more general level, literature increasingly shows that planning has failed to achieve its intended objectives. Urban poverty, low standard of living, environmental pollution and exclusion of citizens from the management of their settlements are observed from various cities of the world, developed and the developing alike. Summarizing the situation in African urban areas, Wekwete concludes:

> The state of African city planning and management in the 1990s is generally depressing as depicted by crumbling infrastructure, unplanned settlements, collapsed governance structures and lack of implementation capacity (Wekwete, 1998).[11]

In addition to the substantive problems of urbanization resulting from and challenging the poor performance of urban planning, the collapse of the Eastern politics, and the refueling of the western democratic pluralism at global level calls for a new planning paradigm.

The dominatory planning based on systematic reasoning no longer holds in a world, which is progressively pursuing the democratic goal.

The driving force behind the new planning paradigm is the belief among theoreticians such as Habermas that, members in societies which cherish democratic ideals, can manage their settlement development issues together while recognizing and respecting their differences. Such a belief provides a basis for the argument that; the plan-led regulatory planning should give way to communicative and collaborative planning where different stakeholders take part in processes of managing change. Planning through debate should therefore replace the old concept where urban planning is understood as a systematic rational thinking carried out by experts through the state bureaucracies. Instead stakeholders should work together. Dreyzek conceptualizes the new planning as a process of deliberation which weighs beliefs, principles and actions under conditions of multiple frames for the interpretation and evaluation of the world (Fischer and Forester 1993).[12] The state of the art of urban planning theory at global level aims at reforming planning practice to adopt the communicative dialogue based on participatory planning practices which facilitate access to the decision making arena to different members of the society and from all levels of government.

Under the new conception, good urban planning practice is supposed to ensure that; diverse people from different societal conditions and cultural communities are encouraged to recognise one another's presence and negotiate their shared consensus(Fischer and Forester 1993).[13] Within the concept of planning through debate, decisions related to for instance, organization of space in cities and how the problem of poverty and unemployment should be reduced concern all citizens as well as their government. Policies and solutions to those problems are therefore adopted, not because they are government directives but because stakeholders have participated in the discourse and are convinced that the policy arrived at through consensus will work. Although at theoretical level there seems to be an agreement as to what is good urban planning, there are however disagreements on how to reach the goal of good planning. On one side there are those who believe that as long as there are rules and regulations stakeholders participation will be free and open. Participating stakeholders will not constraint each other. According to Habermas, as quoted in Flyvbjerg (1998),[14] argumentation ensures that all concerned, in principle take part, freely and equally in a cooperative search for truth, where nothing coerces any one, except the force of the better argument.

Another group argues that, the existence of rules and regulations may not ensure a free and open discourse because, rules only tell you how the game should be played; they do not make the game. Drawing from the Danish case mentioned earlier by Flyvbjerg and Putnam (1993) reporting from a study of two regions in Italy, argue that the belief that the existence of directives and legislation or writing constitutions, as they put it, alters practice is a hypothesis which is hardly supported by historical or empirical studies. Rules and regulations are therefore not sufficient. A society working for more democracy needs to have patience and focus upon more participation, more transparency and more civic reciprocity in public decision-making. Flyvbjerg suggests that:

> ... tradition shows that forms of participation that are practical, committed and ready for conflict provide a superior paradigm of democratic virtue than forms of participation that are discursive, detached, and consensus-dependent, that is rational (Flyvbjerg 1998: 3).[15]

Examining the two positions together, five key principles of good urban planning emerge. Although these principles of good urban planning are mainly developed from urban planning experiences from the capitalist world, given the current socio-economic changes at global level, they have acquired some universal acceptance to the extent that they provide ideal standard, which can inspire practicing planners in Tanzania. In the following sections the principles are outlined and later used to measure and locate the state of the art of Tanzania's urban planning practice.

PRINCIPLES OF GOOD URBAN PLANNING

a) Inclusion Rather than Exclusion

For good urban planning to work, all parties affected or likely to be affected by a given planning issue should be included in the planning and implementation discussions. Once stakeholders are included in the decision making process, they should be assured equal opportunity to present their views and criticize other views. Otherwise stakeholders are encouraged to struggle for possibilities of ensuring that they participate equally.

b) Suitable Legislation and Institutional Set-up

Another principle for good planning is a legislative and institutional framework, which neutralises existing power differences among participants. Power differences between stakeholders have no effect on the process of consensus building. Healey describes that situation of equal opportunities to discuss as:

> ... intercommunicative planning involving respectful discussion within and between discursive communities, respect implying recognizing, valuing, listening to, and searching for translative possibilities between different discourse communities (Healey, P. 1993: 247).[16]

c) Local Negotiation Fora

Creation and facilitation of local negotiation fora is one of the key principles for good planning. Communicative planning should construct an arena within which different stakeholders can formulate programs, and identify and mediate conflicts. To illustrate this principle we draw an example from South Africa, as reported by Halfani:

> After four and a half decades of violent confrontations at local level, the contending factions in South Africa initiated negotiations at the local level. These interactions resulted in the creation of local negotiating fora... These integrated socio-political fora established a foundation for inclusiveness, representativeness. legitimacy and the engagement of civil society in the process of urban governance (Halfani, M. 1998: 27).[17]

In my view, the most important role of urban planning in a society aspiring for open and participatory planning should be to facilitate the creation and the working of local negotiating fora.

d) Agreements as Temporary Products?

Within collaborative planning, negotiation lead to outcomes, which are, regarded as agreements, either in form of framework, comprehensive plan or even in form of a process of dealing with conflicts as and when they occur. Such agreements should not be taken as static. They are temporary products which can and should be changed if the parties to that so wish.

e) Collaborative Planning is not a Hit and Run Business

The new concept of urban planning requires changes in the formal and informal institutions so as to contribute to changes in the decision making practice. According to Flyvbjerg, institutional change typically moves much more slowly and circuitously than is often assumed by legal writers and institutional reformists (Flyvbjerg, 1998: 235).[18] This principle consoles societies pursuing the good governance objective that, they have to be patient. To make democracy work in practice may take centuries. Thus building up institutions which will produce good urban planning practice might be depressing for those who are impatient. Participatory process take long to deliver but once the practice is well established, it works better and it is sustainable.

The five principles outlined here which try to summarize the requirements for good urban planning practice also form a vision for improved planning for societies aspiring for more democratic practices. We shall now part company with the western theoreticians for a while and bring the discussion back to Tanzania. In the following section we shall relate the Tanzanian urban planning practice with these principles. We shall provide evidence on how close is the Tanzanian urban planning to the principles, and in that way we shall be outlining the state of the art of urban planning in Tanzania.

URBAN PLANNING PRACTICE IN TANZANIA

Historical Overview

Arab traders introduced the plan-led regulatory mechanism by which the government or community may influence urban development in Tanzania. In 1866 Sultan Said Bin Majid made and implemented a plan for the development of the town of Dar es Salaam. Prior to that, tribal or clan leaders managed their settlements on the basis of cultural norms. When the Germans took over Tanzania, (then Tanganyika) from the Arabs, they made a plan for the harbour and the central area of Dar es Salaam. Mandated for the state to carry out urban planning, i.e., to influence land use and land development was made clear and legalized by the Germans through the 1889 declaration that "all land in Tanzania whether occupied or not is *crown land* vested under the German Emperor (James 1971)."[19] According to Okoth-Ogendo:

> The declaration extinguished the superiority title of landowners, both individually and collectively. The ordinance also gave the government powers to resume possession of land granted under any right or interest, should *public purposes or interest* so require, and to prescribe conditions under which rights granted by it could be enjoyed.... (Okoth Ogendo, 1995: 4-5). [20] (*emphasis added*)

Through that Ordinance the Germans managed to alienate land and to control land use. That declaration was later enacted into law in 1923 by the British, and became the Land Ordinance which, has with minor revisions guided land administration in Tanzania to date. In Tanzania, land ownership is vested in the President as the custodian on behalf of the citizens. The Land Ordinance recognises two types of rights to land: government granted rights of occupancy under a leasehold system for 33, 66 and 99 years, and deemed rights of occupancy for land occupied under the customary land tenure system. In the process of preparation or adoption of a land use plan or a master plan, the 1956 Town Planning Ordinance, which is discussed in the following section, requires that the Minister, through local authorities, pay compensation to the properties to be adversely affected. In addition, official transactions affecting undeveloped land are assumed to have zero exchange value, implying that in case of displacement of land rights, the holder of land rights is, in accordance to the Land Ordinance, only entitled to compensation of what in the law is referred to as "unexhausted improvement," such as buildings and or crops. In law, 'unexhausted improvement' is defined as anything or any quality permanently to the land directly resulting from the expenditure of capital or labour by an 'occupier.' Although crops are considered for compensation, the rates are normally outdated thus making assessment based therein far below the market value.[21] The criteria used for assessing compensation whereby ordinary cultivation is not taken into account, is the major source of many conflicts in land development in urban areas, particularly in planning and land development in peri-urban areas. In future conflicts in land administration and in land rights may however be minimized once the Land law based on the 1995 National Land Policy (URT, 1995)[22] becomes operational.

One of the important provisions made by the new Land Law with respect to land use planning and urban land development is the recognition that land has value and that in case of land acquisition by the government, compensation based on market value should be effected fairly and promptly. The more clear allocation of functions and responsibilities among government, private sector and communities will also make a significant contribution towards minimizing conflicts over urban lands. However more time is needed to experience the real contribution of the new land law for improved management of urban land.

The 1956 Town and Country Planning Ordinance (as revised in 1961)

The British who took over Tanzania from the Germans put urban planning practice into law in Tanzania. The 1936 Town Development Control Ordinance outlined plan preparation and implementation procedures. On the basis of the ordinance, the British colonial administration prepared urban planning schemes for various towns including Dar es Salaam. Thirteen years later, the Town Development Control Ordinance was found inadequate to address urban development issues of that time. In 1949 the British drafted another ordinance drawing heavily from the British Town and Country Planning Act, of 1947. The new ordinance for Tanzania, The Town and Country Planning Ordinance, became law in 1956. After independence in 1961 the ordinance was subsequently amended and adopted as the principal legislation for urban planning in Tanzania. It is in operation to date. As summarised in Table 5.1, the main focus of the ordinance is, like the 1936 Town Development Control Ordinance, which it replaced, procedures for preparation and approval of general and detailed planning schemes.

Table 5.1: *Plan Preparation, Approval Procedures and Authority* (as provided for in the Town and Country Planning Ordinance, 1956 and as revised in 1961)

Step and Activity	*Authority*	*Time span*
Preparations: Declaration of a town or part of as a planning area.	Minister in consultation with the local authority	Not specified
Preparation of a general planning scheme	Minister or local authority (Area Planning Committee)	Not specified
Approving general planning scheme for publication	Minister	Not specified
Publication of the plan and deposition of the plan for public inspection.	Committee	Not specified
Inspection of the plan by the public and receiving objections	Minister, Area Planning Committee, and the public	Two months
Depositing copies of the objections for further public scrutiny and representations.	Area Planning Committee and the public	One month
Submission of the scheme plus objections to the minister for decision.	Minister and Area Planning Committee	One month
Approval of general planning scheme with or without modifications.	Minister	Not specified
Decide to prepare a detailed planning scheme for a planning area	Minister or local authority	Not specified
Inform land owners of intention to prepare detailed scheme. Landowners may prepare their own scheme.	Minister and Area Planning Committee	Six months
Preparation of detailed planning scheme for a planning area	Minister and Area Planning Committee	Not specified
Approval for publication of a detailed planning scheme	Minister	Not specified
Publication of a detailed planning scheme and deposition for public inspection.	Minister and Area Planning Committee	Not specified
Receiving objections.	Minister and Area Planning Committee	Two months
Depositing copies of objections for further public scrutiny and representations.	Area Planning Committee and the public	One month
Submission of scheme plus objections to the Minister for consideration and decision	Area Planning Committee	One month to
Approval of detailed scheme	Minister	Not specified

In addition to the procedures, issues to be dealt with by town and country planning authorities include:

a) Use of land and buildings;
b) Intensity of use of land and occupancy rate;
c) Size, form and construction materials of buildings;
d) Siting of buildings, alignment of and reservation of land for roads and other physical infrastructure; and
e) Preservation of natural and man made features, as well as regulating and controlling disposal of refuse and pollutants.

According to the pattern of task allocation, the main actor and also the authority in plan preparation and approval is the Minister for Lands and Human Settlements Development. The same minister has authority to initiate a plan- making process through declaration of planning area, and sees to it that legal procedures are followed and when satisfied by his or her own activities or those of his or her officers s/he approves the plan.

It is also clear from the Table 5.1 that time is not specified for tasks to be undertaken by the minister. Time is specified for tasks involving the public. In terms of the time span it is thus difficult for the public to legally hold the minister responsible. There are cases of master plans, which have been under preparation for more than ten years. In the same spirit of limiting public participation in the plan preparation process the Ordinance allows the minister to disregard objections raised by the public, as s/he may approve a scheme with or without modifications. In other words it is not mandatory that objections are considered and incorporated into the plan.

Commenting on the relationship between the government and the public on the basis of provisions made by the Ordinance, Kironde relates the high centralization of urban planning powers to the central government with the centralised and bureaucratic system of governance under the British colonial administration "where the Governor's consent had to be sought even for such a trivial decision as cutting of trees in townships (Kironde, J. M. 1995)."[23]

Actors and How They are Organised

The organization of actors responsible for urban planning follows the government administrative structure. Although it is the minister for Lands and Human Settlement Development who has the overall mandate and authority for carrying out urban planning, the actual execution of planning functions is the responsibility of the directorate of human settlement in the Ministry as shown in Figure 5.1, is organised into five sections which are:

a) Urban Development Policy and legislation;
b) Planning of Prime Areas and Records Keeping;
c) Housing Policy;
d) Urban Strategic Development Planning; and
e) Revolving Housing Fund.

At the local level town planners working under the respective local authorities represent the directorate. Following the 1997 Regional Administration Act, town planners earlier posted in the regional offices, have been posted to local authorities as part of the decentralisation and civil service reforms. Although the decentralisation process is being implemented, for instance by shifting manpower to local authorities, the government is yet to clarify the type and form of urban planning relationships between the ministry, the regional secretariat established through the 1997 Regional Administration Act, and the rural as well as urban local authorities. Without clarifying that important institutional and operational relationship it is difficult to really decentralise urban planning practice in Tanzania. Thus among the agenda for reforming and improving urban planning in Tanzania is to chart out an institutional framework which clearly defines responsibilities and tasks among the different levels of government actors. However, in order to have a deeper understanding of how the legislation and the actors work, we now examine the practice of urban planning in details. We start by the main activity of the Tanzanian urban planning enterprise, which is to make plans.

Production of Urban Master Plans

By and large, the urban planning system in Tanzania is guided by a belief that through preparation, approval and implementation of master plans, it is possible for the government to influence urban development. This belief in plan-led regulatory framework has its origin and theoretical support from the 1950s conceptualisation of planning as design, following the well known *survey-analysis-plan* process as was conceived by one of the fathers of British planning, Sir Patrick Geddes, more than seventy years ago. According to Healey, during that period, town planning was conceived as:

> Centrally focused upon the technical processes of preparing plans to remedy and improve the conditions of towns. Plan preparation involved survey of existing conditions, certain calculations to establish the population size and scale of activities to be accommodated in a town and the translation of these quantities into a physical arrangement through the use of various town design principles described by Keeble as 'Planning theories' (Healy, 1985).[24]

A close examination of the process and contents of urban master plans shows the dominance of the survey-analysis-plan process. Planners from the directorate in the Ministry of Lands and Human Settlement Development or private consultants collect data through various survey methods, analyse the data and produce a plan. Following that approach a master plan for Dar es Salaam was prepared by a Canadian private consulting firm in 1968. Later on, another foreign firm prepared a master plan for Arusha town in 1970. As part of the implementation of the 1970s growth pole concept in Tanzania, master plans were also prepared for Tanga, Moshi, Morogoro, Mbeya and Tabora towns. Since then the main activity in the directorate of town planning is preparation of master plans.

At the 1997 Planners Annual Meeting[25] the production of plans in Tanzania was evaluated. The key paper given by the then Director for Urban Planning, Professor Hayuma, provided the state of the art of master plan preparation in Tanzania:

> Out of the 20 regional headquarters towns, the master plans of six towns: Morogoro, Moshi, Tabora, Shinyanga, Bukoba, and Mwanza need to be reviewed (The law requires that plans are reviewed five years after approval). Work on the first four towns has started, but requires at least Tshs. 44 million to complete. Out of the 76 district headquarter towns, it is only 48 towns that have valid master plans or interim land use plans. The master plans/interim land use plans of 30 towns require review. A total of 28 district headquarters towns have no master plans or interim land use plans. Individual town planning drawings guides their development.... Where there are suitable base maps on scale of 1:2500 there are no problems with the preparation of detailed town planning drawings. Town planning offices in the districts, regions and urban councils are satisfactorily executing this work. Plans are submitted to the director of urban development for approval after they have been endorsed by the respective urban planning committees as required by law (Hayuma, A. 1997: 6).[26]

This statement suggests that the issue undermining the performance of urban planning is not the approach but the lack of financial capacity to finance the plan making process. The Minister also considers insufficient amount of funds a constraint. For instance, the Minister's 1998/99[27] Budget Speech informed the Members of Parliament that, because of lack of funds and poor communication, the ministry could not accomplish the preparation of master plans for Tabora and Shinyanga regional headquarters towns, and for Bunda district town (Minister's Budget Speech, 1998: 40). The government does not allocate sufficient funds to the Ministry responsible for preparation of plans.

Hayuma associates the failure of planning to keep ahead of development not only to lack of funds but also:

> Because planners *lack power* to plan the towns according to *their rational schemes* and that the attitudes of the people and the local authorities seem to be more inclined to individualism and opportunism and expediency rather than civic consciousness (Hayuma, A. 1997: 4).[28] (*emphasis mine*)

It is depressing to find out that practicing planners, especially those at the Head Office who are expected to advance the practice, are still relying on the 1950s planning philosophy and tools. None of the eight papers presented during the 1997 annual meeting touched upon the current theories in governance and urban planning. None of the new concepts of planning seem to be informing the planners' discourse. The majority of the Tanzanian practicing planners, especially those working for the government - central or local - still believe or conceptualise planning as plan production. They believe in top-down planning where the main actor is the central government and the main activity is production of plans for their own sake. In my view such position towards urban planning retards the current government reform measures towards decentralized and more democratic governance. We support this view by case stories in the following sections.

In addition to holding back the reforms, the planner's conceptualisation of planning contradicts the spirit of the National Human Settlements Development Policy which is under preparation by the Ministry of Lands and Human Settlements Development. For example, one of the policy objectives in the draft report is to promote the participation of the private and popular sectors in planning, development and management of human settlements (URT, 1998: 21).[29] There are also policy statements, which show government intention to facilitate provision of and improved services and infrastructure through supporting CBOs and NGOs. These policy intentions which are in line with the new governance and urban planning paradigm are thus calling for more effective reforms in the planning system, to recognize and incorporate CBOs, NGOs, and the private sector in managing urban development.

Unless town planners, who are among the key implementers of the reforms, change their attitude from centralized exclusionary planning practices, movement towards more democratic and accountable planning practice is likely to remain a dream.

LOCAL LEVEL URBAN PLANNING PRACTICE

Collaborative Planning in the City of Dar es Salaam

Another area where we can also see the state of the art of urban planning is at local council level. In the early 1990s collaborative planning approach was initiated in the city of Dar es Salaam as a strategy to cope

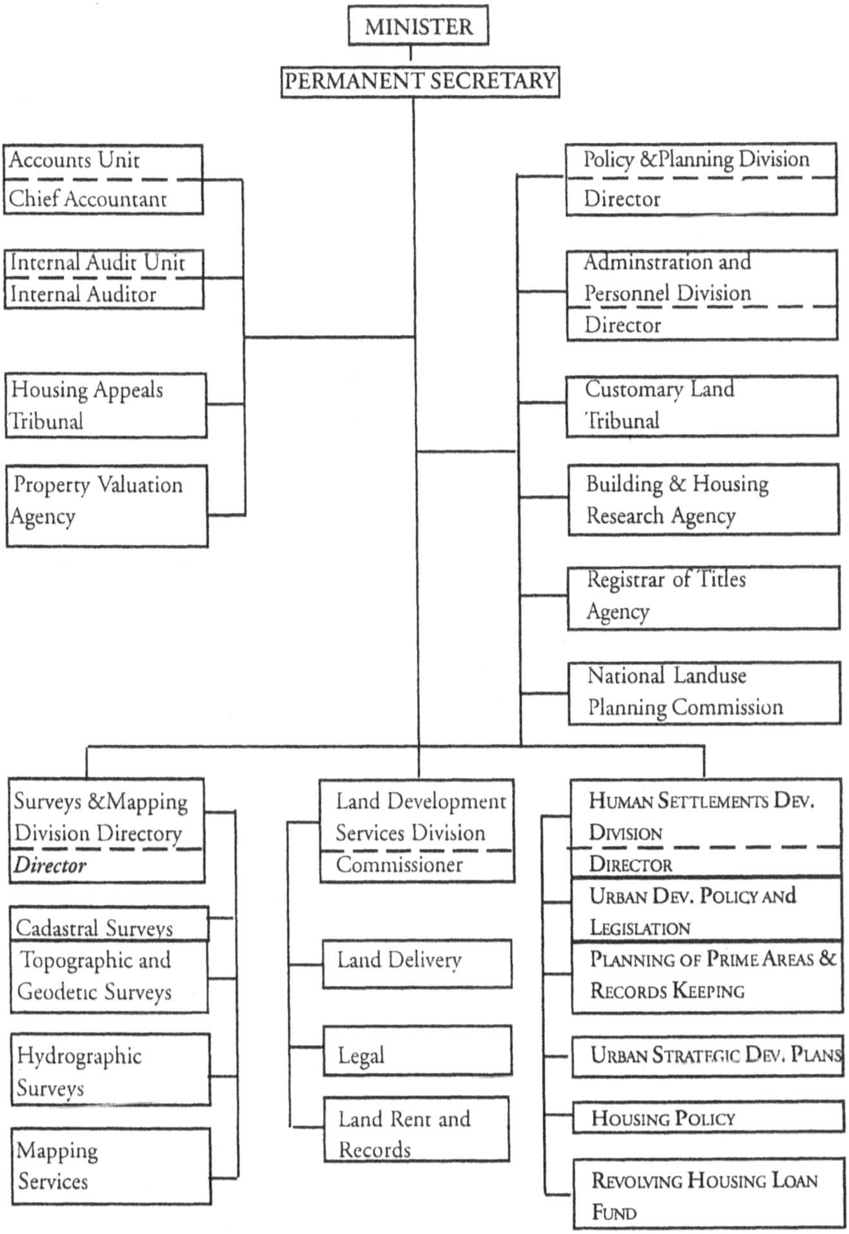

Figure 5.1: *The Human Settlements Development Division in the Ministry of Lands and Human Settlements Development.*

with urban management crisis resulting from the poor performance of the centralised urban planning. The urban population has been increasing at a rate of 10% per annum between 1980 and 1997 (URT, 1998: 5).[30] According to the Strategic Urban Development Plan for Dar es-Salaam, the population of the city is growing at an average rate of 8 per cent per annum making Dar es Salaam one of the fastest growing cities in Sub-Saharan Africa (Dar es Salaam City Commission, 1998: 1).[31]

In order to accommodate the three million people living in Dar es Salaam, (or 10 per cent of the national urban population) the city residential areas expanded so rapidly that the available urban planning mechanisms and land allocation systems failed to cope. One of the consequences is the increase in the number and size of unplanned and poorly serviced settlements located in flood plains and in steep slopes. Storm water drainage and sanitation are critical environmental problems in the city. Because of limited employment opportunities in the formal commercial and industrial sectors the growing city population opted for small scale trading and vending activities, in most of the streets in the city centre. Traffic congestion and piles of uncollected garbage added to the issues facing the management of the city in general and urban planners in particular.

In response to that situation the government, through the Ministry of Lands requested for support from the UNDP to review the 1979 Dar es Salaam Master Plan, on the assumption that the problems facing the City could be addressed through the master plan. The UNCHS-HABITAT advised the City Council to adopt a more effective urban planning approach, which focuses on building the capacity of the City in planning and managing growth and development in a sustainable way. It was thus agreed by the government that the new planning approach based on Habitat's Sustainable Cities Program be initiated in the city. A project, known as Sustainable Dar es Salaam Project (SDP) was therefore started in 1992, aimed at improving the capacity of the government institutions to coordinate different stakeholders for sustainable urban development. The Sustainable Dar es Salaam Project is conceptually hinged on the Sustainable Cities Program (SCP) which is a global initiative to implement Agenda 21 on the environment. Specifically the SCP aims at improving environmental planning and management capacities through reforming urban planning approaches. The approach adopted by the SCP is however in line with the new conceptualisation of planning as a collaborative activity. In Tanzania, the Sustainable Dar es Salaam Project is guided by the principle of:

> Commitment to participation in urban environmental planning and management of all interested groups of the city: the public sector (central and local), the private sector

> (formal and informal), community and NGOs sectors... and the principle of reliance upon structured collaboration and negotiation... which can bring together the relevant stakeholders to focus on agreed priority issues which are in turn continuously re-assessed (UNDP/URT, 1995: 5).[32]

The new approach to urban planning introduced in Tanzania through the Sustainable Dar es Salaam Project, offers opportunity to reform urban planning in the direction of the good urban planning principles outlined above. The Sustainable Dar es Salaam Project was positively evaluated in 1995 and it was decided to introduce the approach to the nine municipalities in Tanzania (For more discussion on the SDP see for example, Halla, 1994).[33]

Of the nine municipalities, Moshi, Tanga, Mwanza, Tabora and Dodoma, seem to be on the forefront in adopting and carrying out urban management in accordance with the principles of the new paradigm. It is interesting to note that, while the municipalities are keen to improve the performance of urban planning, and to shift from the master plan-led planning approach to more inclusive planning based on debate and negotiations, the Ministry of Lands is still busy with the production of master plans. For instance the Ministry of Lands is finalising the preparation of Moshi master plan. In general, reforms towards more open, decentralized planning are taking a rather slow pace in the Ministry. Those who wish to see faster rate of reforms are however consoled by the changes taking place at local council level and within communities.

In the conception of planning as a communicative enterprise, it is a prerequisite that communities take part in planning activities. Cases of communities attempting to influence planning decisions and wishing to participate in managing their urban environments are plenty in Tanzania. Because we recognise the role played by communities in managing their settlements, it is important that cases of community initiatives are included in this discussion of the state of the art of urban planning. We have thus selected cases from Moshi municipality in Kilimanjaro and from Makambako, a small town in Iringa, Southern Tanzania. We start with the story from Kilimanjaro.

Moshi: Collaborative Planning Turned Down by Municipal Planners

Longuo is one of the residential neighborhoods in the Moshi municipality in Kilimanjaro Region. The settlement is located between the Kilimanjaro Christian Medical Centre and the Police Training School, thus making it attractive for rental accommodation. In a survey carried out in 1994 for a research on planning practice in Tanzania, Nnkya reports that the settlement which is only 36 hectares accommodates 272 housing units and about 3600 people (Nkya, 1996).[34]

By Tanzanian urban planning standards, Longuo is unplanned. It is historically a settlement, which developed spontaneously from a purely rural and agricultural base to an urban residential area, mainly serving tenants from the town and the nearby institutions. Planners however considered the settlements as rural and in the 1974-1994 Moshi Town Master Plan, Longuo was zoned for public institutions. Had the master plan proposal been implemented people living in the settlement would have been required to shift to other areas after compensating the property owners.

Realising that the proposal in the master plan could not be implemented for various reasons including lack of money to pay compensation, the Regional Town Planning Officer in Moshi prepared an upgrading plan for the area. This meant that the Regional Town Planning Officer considered it worthwhile to recognise the area as a residential neighbourhood and thus introduce urban planning standards through a detailed plan - the upgrading plan. In general, a settlement upgrading plan which, according to the 1956 Town and Country Planning Ordinance, may be referred to as a detailed planning (or regularisation) scheme, means the residential development would be officially recognized and thus the proposal in the master plan would no longer be honoured. However, land parcels, roads and sites for community facilities have to be re-organized in accordance with urban planning concepts and standards.[35] The Longuo upgrading plan was in 1985 approved by the Director for Town Planning for implementation. Again, the plan was not implemented for similar reasons to those, which constrained the realization of the master plan proposal.

Findings from the research (see Nkya, 1996)[36] show that local leaders in Longuo were also not happy with the spontaneous nature of the physical development in the area. They decided to seek advice from the municipality's urban planning unit. In 1989, the local councilor organized a ward development committee meeting to which a town planner and a land development officer from Moshi municipality were also invited. During the meeting the councilor expressed concern over the haphazard nature of development in Longuo. The local leader was concerned with the real and potential problems created by the high rate of land subdivision and land selling triggered off by fear among residents that the municipality was planning to acquire un-built parcels of land in the settlement. He argued that, if that trend was left to continue, building densities would be very high. Some compounds will not have space for building pit latrines and some roads and access paths will be blocked and may disappear. In response the two experts informed the residents on the procedures for obtaining *certificates of title* for their parcels of land.

After the meeting, 14 landowners wrote to the council requesting for a permit to have their *'shambas'* surveyed. The councilor for Longuo who chaired the above mentioned meeting, and who was also among the 14 land owners and the main actor in trying to facilitate the survey argued thus:

> My first agenda as a councilor for Longuo (after being elected to office in 1988) was to convince the municipality to accept Longuo settlement as it is ?i.e. as residential neighborhood - because the government had no money for compensation, and the municipality had no land to resettle the would be displacees. The municipality accepted that request. I then started to educate people that, eventually, the municipality would prepare a detailed plan for the area because the existing settlement pattern was unacceptable to them. If they do so, houses would be demolished.... I told them the strategy would be: first, to carry out survey of the area to show the municipal authority how we are settled. That picture of our settlement will then be taken to the municipality to see whether there will be a need to resettle any of us. This (approach), I argued, would be far better than to let the municipality resettle all of us.
>
> Having done that, I convened a meeting of the local leaders in Longuo, the party chairman and ten cell leaders. I told them that our area being part of the town, *we could no longer continue to live like goats in a shed.* We needed to have a plan, which is recognized by the government. We then resolved that each ten-cell leader should educate his/her people on that need. This task took a whole year. The next step was to find a land surveyor to carry out the survey. I met one who agreed to do the work for Tshs. 1,000 per land owner, because it was not a government assignment(Nkya, 1996: 126-127).[37] *(emphasis added)*

Although, judging from the councilor's words, landowners were ready to collaborate with the municipality and finance the land registration process, the planners in the municipality were not ready to coordinate the residents. When interviewed by the researcher as to why they turned down the request from the community leaders the planners responded:

> We (the planning department) never like to get involved in coordinating people and collecting money from them. [This is because] in case of a problem it would be our responsibility. Let an outsider organise them, so that in case of a problem we will not be accountable...We also use local leaders such as Sub-ward chairpersons or Councilors, to organize people...normally our work is not to mobilise people. Once a plan for an area has been prepared, we ask the local leadership to inform their people about it and urge them to seek survey permission (Nkya, 1996: 2).[38]

The surveyor took over the planner's role. He collected the money from the landowners and carried out the survey. In addition to dermacating plot boundaries, the surveyor working together with the landowners, agreed and mapped new road alignments and widths. He also advised the landowners to subdivide the large parcels of land into housing plots in accordance to the municipal urban planning standards. Where a compound lacked access the surveyor also advised the landowners to negotiate for the same. In 1991 the surveyor accomplished the survey plan for Longuo, which the residents regard as the plan for Longuo although it is not approved by the respective authorities in Moshi municipality and in the Ministry of Lands.

As from 1996, following the introduction of the Sustainable Cities Program approach in Moshi Municipality, there are attempts to reform the planning practice towards more collaboration with other stakeholders, like the people in Longuo. It is therefore expected that when the new approach is well developed and adopted, we shall be able to get different planning practice stories from Moshi.

The Moshi case about what actions can be triggered off at local level by a master plan is not an isolated one. It is rather a common phenomenon in urban Tanzania, especially in residential areas where landowners feel insecure about their land rights or are threatened by interventions from government town planners. Similar cases or incidences have been reported from Makongo and Changanyikeni neighbourhoods of Dar es Salaam and parts of Arusha municipality. However, for purposes of the subject matter of this chapter - the state of the art of urban planning in Tanzania - the following issues from the Moshi story seem to be pertinent. Communities and their leaders are aware of the need for planning and they believe that they can be supported by experts from their local authority:

a) Communities are ready to collaborate and work together with experts. The opposite is true for government planners because they perceive planning as a monopoly of the government;
b) Communities are ready to pay for services rendered hence their requests are basically technical and not financial;
c) Government planners are also not keen to coordinate resident's efforts because they do not like to be accountable.

The lessons from Kilimanjaro, probably the most important argument which is illustrated here, is that in attempting to reform planning towards a more open and collaborative practice more efforts and resources should be directed towards ensuring appropriate education for planners particularly in changing their attitudes.

Makambako: Joint Management and Service Provision

By Tanzanian settlement development policies and standards, Makambako settlement, with a population of 26, 888 people (URT, 1998)[39] is regarded as a trading centre. It is not yet recognized as a township authority. Its respective district council of Njombe in Iringa Region, southern Tanzania, thus administers it. Administratively Makambako is made up of Ubena, Mwembetogwa and Mjimwema villages, which together make up Makambako ward. The center is both a ward and a divisional headquarters. In addition to the administrative functions, the center is transversed by the Tanzania - Zambia highway and the Makambako Songea trunk road. The center is also one of the main stations along the Tanzania-Zambia railway. With such local and national level activities Makambako has been urbanizing fairly fast compared to other settlements in Njombe district. Its rate of physical development has been rather fast compelling the district council to develop strategies in which existing local capacities in the centre are organised and directed towards improving the management and provision of services in the centre. In a study carried out in 1993 by the author of this chapter (Lerise, 1993)[40] the following institutional collaboration for improved settlement management and service prevision was observed.

Land development particularly in the housing and commercial areas was the main issue in Makambako in 1993. To ensure ordered development, Njombe District Lands Development Officer in collaboration with ward development committee at Makambako agreed to share the responsibility of controlling the construction process in the center. That joint planning and management worked in the following way: For construction to be allowed a developer has to obtain a building permit from the district office. To ensure that the building includes health considerations, and that it was structurally sound, the district health officer and the district engineer scrutinize permits. The land development officer and the town planner, all from the district office confirm issues of land rights and land use considerations. During construction the Ward Secretary assists in ensuring that developers build according to the permit issued. The ward development committee handled land conflicts.

Another area where the district collaborates with the local level institutions in Makambako is in the management of solid waste within the centre. Because of increased trading activities and business at the main bus stand, individual traders in the centre could not manage the solid waste generated. The district health officer organised the traders into an association. Among the responsibilities of the association is management of solid waste at the market. The Traders Association then decided to engage a local group of hand-pulled cart owners in collecting and disposing of the solid waste.

The Cart Owners Association was then contracted to carry the waste generated at the market and at the bus stand and dispose of it. During the survey in 1993, it was observed that solid waste was collected and disposed of at a site outside the built-up area of the centre, leaving the market and the bus stand comparatively clean.

However, apart from the disposal site being located close to residential development, it was also noted that the Cart Owners Association was only concerned with dumping the waste. They did nothing to manage the disposal site. Keeping the commercial part of the town clean was not reconciled with the needs of other residents around the dumping site. Neither was it related to the town-wide objective of keeping the overall settlement clean. As more waste is likely to be generated, the District Council may have to assist the traders and the cart owners to chart out ways to manage the dumpsite and also to identify and acquire a new disposal site. That assistance may require an overall plan of solid waste management for the trading centre.

The collaboration observed in the small town of Makambako in 1993 seems to be a common strategy adopted by district (rural) as well as urban councils in managing waste. From 1994 and later in 1996, Dar es Salaam City Commission for instance collaborates with five private companies in managing solid waste generated in the city. Collaboration in improving service provision is also found in other public services such as parking, toilets, construction of drainage etc. For further information on the privatization of municipal service provision in Tanzania, see for example, Lerise (1998).[41]

Since collaboration between local councils and the private sector in improving service provision in urban settlements seem to be the norm rather than the exception, service provision is no longer a monopoly of city planners. Planning for service provision is thus increasingly becoming a joint activity involving residents, private companies and local authorities. It is also worth noting that such collaborations work better in areas like markets and bus stands where the main users can organise, and have, organised themselves. The capacity of civic groups to organise themselves, or that of town planners to organise service users, is an important factor for the new approach to improved service provision. That is one of the challenges facing Tanzanian urban planners. Although there has not been comprehensive studies yet on the implication of the new arrangement in service provision on poverty reduction in Tanzania, general impressions suggest that decentralisation and privatisation of service provision have a high potential in poverty reduction. It is thus an approach, which should be encouraged through urban planning initiatives not only for environmental reasons but also employment and income generation purposes.

The fact that the number of CBOs and NGOs in urban Tanzania is increasing rather fast, provides a more fertile ground for improved service provision and for collaborative planning. The challenge facing practicing planners and urban managers is how to facilitate the increasing number of grassroots organisations to participate and contribute towards environmental management, sustainable resource utilisation, poverty reduction and in general to take part in the overall governance process.

CONCLUSION

After the detailed discussion above we are now at a point where we can locate the Tanzanian urban planning practice in the continuum of global reforms. Using the five principles and the case studies outlined above, it is clear that the Tanzanian urban planning practice is far from the current global vision of urban planning. The practice is still guided by the 1950s technical rationality and the centralised planning carried out through state bureaucracy. Efforts to reform the system to fit the national and local processes towards more democratic planning are underway but have achieved very limited results particularly at the top, i.e. the central government level.

The stories from Dar es Salaam, Kilimanjaro and Iringa point to an emerging trend of creating local negotiation fora. Some of these fora are recognised by the central and local governments; others are not. In my view recognition of such fora and their facilitation will also reduce the burden to central government institutions in dealing with local issues and conflicts. That may also make available more central government resources for other issues, which are of national importance. On the other hand, planners and decision-makers in government should honor agreements reached in the negotiation fora. They should therefore accept the idea that plans are agreements and should involve relevant actors making such agreements.

Exclusionary practices are still common in public decision making arena whether at local or at national level. Important decisions such as land use changes or land acquisition are still being made without involving land users and landowners. Collaborative planning is yet to be adequately adopted as a tool for public decision making in urban Tanzania. The current exclusionary practices, which seem to be encouraged by the legislative and institutional framework, do not enhance and stand for decentralised and participatory decision making practices. It is likely that the government sees the need to invest in reviewing the current institutional set up of planning. In that case, more research focusing on planning practice will be required to generate more practice stories, which are needed in the reforms.

The poor state of the art regarding urban planning practice knowledge grounded in Tanzanian context, is a result of insufficient research in planning practice. The limited research in urban planning in Tanzania is mainly focused on the substantive issues rather than the actual process and practice of planning. There is therefore knowledge on urbanization, rural-urban migration, service provision, growth of settlements and land administration. Recent literature covering these issues include Kironde, (1994);[42] Kombe, (1995)[43] Lugalla, (1995).[44]

It is only recently that planning research has started to produce cases which evaluate planning practice and thus more useful for improving the skills of planners and as inputs into reforming the planning process. Most of the literature has been extensively cited and used in this chapter. More effort is needed - especially in this period of planning reform - in searching for practice stories. It is therefore important for researchers to chart out strategies and policies as well as appropriate actions which will ensure that knowledge on planning practice, especially the best practices, is continuously produced and made available to decision makers as well as practicing planners.

END NOTES

1 United Republic of Tanzania: *National Report on Human Settlements Development in Tanzania.* Prepared for HABITAT II, March 1996: 38.)

2 United Republic of Tanzania and the Ministry of Foreign Affairs/DANIDA: *Preparation of the Strategic Plan for the Restructuring of the City Council of Dar es Salaam.* COWI Consult, 1997: 132)

3 Halfani, M., *State of Knowledge and the challenges of operationalizing governance in Tanzanian type of societies.* An unpublished paper presented to a workshop on Institutional Reforms and its Impact on Urban Governance in Tanzania. April, 1997. Dar es Salaam.

4 *Maclean's Weekly* Feb. 7, 1994 quoted by Qadeer, M.A. "Pluralistic Planning for Multi-

5 Devas, N. and Rakodi, C., (eds.) *Managing Fast Growing Cities: New Approaches to Urban Planning and Management in the Developing World.* Longman Group UK LTD, 1993, Essex.

6 Friedmann, J., *Planning in the Public Domain, From Knowledge to Action,* Princeton University Press, 1987, New Jersey. p. 28

7 Paris, C., *Critical Readings in Planning Theory,* Pergamon Press, Oxford, 1982.

8 Healey, P. McDougall and Thomas, M. J., *Planning Theory Prospects for the 1980s,* Pergamon Press, Oxford, 1982.

9 Bryson, J. M. and Crosby, B. C., *Leadership for the Common Good Tackling Public Problems in a shared-power world.* Jossey - Bass Publishers, San Francisco, 1992.

10 Flyvbjerg, B., 'Empowering Civil Society: Habermas, Foucault and the Question of Conflict." Paper presented to a symposium - University of California, Los Angeles, April 1998. p. 5.

11 Wekwete, K. H., "Urban Poverty Reduction, Towards Mainstreaming and Implementing Urban Poverty Reduction Action Plans in Sub-Saharan Africa." Draft background paper for the Africa Regional Workshop, Nairobi June 1998.

12 Fischer F. and Forester J., (eds.). *The Argumentative Turn in Policy Analysis and Planning,* Duke University Press, London 1993, p. 214.

13 Fischer F, and Forester, J. *Ibid.* p. 244.

14 Habermas, "Moral Consciousness and Communicative Action," as quoted in Flyvbjerg, *Rationality and Power: Democracy in Practice.* University of Chicago Press, Chicago 1998: 3.

15 Flyvbjerg, B., *Rationality and Power: Democracy in Practice.* University of Chicago Press, Chicago p, 236.

16 Healey, quoting John Forester and Geertz, in Fischer F. and Forester J., (eds.). *The Argumentative Turn in Policy Analysis and Planning,* Duke University Press, London 1993: 247.

17 Halfani, M. "The Challenge of Urban Governance in Africa, Institutional Change and Knowledge Gaps in Swilling, M., (ed.) *Governing Africa's Cities* Witwatersrand University Press, Johannesburg, 1998, p. 27

18 Flyvbjerg B., 1998, *op. cit.*, p. 235.

19 James, R. *Land Tenure and Policy in Tanzania,* East African Literature Bureau, Dar es Salaam, 1971

20 Okoth-Ogendo H.W.O. "Final Report on Reform of Land Tenure and Land-use Legislation in Tanzania." Technical Annex 1. Technical Support Services Project URT/94/02T, FAO, Rome (1995) p. 4-5

21 Valuation for compensation is provided for under the Land Acquisition Act of 1967. The Ministry of Agriculture issues crop compensation rates, but they are not regularly updated. The current (1998) crop compensation schedule was issued in 1992.

22 The United Republic of Tanzania, Ministry of Lands, Housing and Urban Development, *National Land Policy.* 1995.

23 Kironde, J. M. *The Evolution of the Land Use Structure of Dar Es Salaam 1890-1990.* An unpublished *Ph.D.* Thesis, University of Nairobi. 1995, p. 239

24 Healey, P. "Whatever Happened to Methodology in Land Use Planning?" *Ekistics* 311, 1985.

25 It is a common practice for the directorate of urban planning to organize an annual event in which practicing planners meet and evaluate their achievements and discuss their problems. The 1997 annual meeting was held in Dar es Salaam from 10 to 11, April.

26 Hayuma, A., M., "Town Planning Practice in Tanzania and Development of Sustainable Human Settlements." Unpublished paper presented at the Annual Meeting of Town Planners in Tanzania, Dar es Salaam, April 1997.

27 Jamhuri ya Muungano wa Tanzania, *Hotuba ya Waziri wa Ardhi na Maendeleo ya Makazi, Mheshimiwa Gideon Cheyo (MB) Akiwasilisha Bungeni Makadirio ya Fedha kwa Mwaka 1998/99.* Government Printers, Dar Es Salaam, 1998. p. 39-40

28 Hayuma, A. M., 1997, *op. cit.*, p.4

29 United Republic of Tanzania, "National Human Settlement Development Draft Policy." Ministry of Lands and Human Settlements Development 1998.

[30] URT, 1998. *Ibid.*

[32] "Strategic Urban Development Plan for Dar es Salaam, Part 1 Volume 1: City Growth/ Expansion/Vision Pattern." First Draft, December 1998.

[33] United Nations Development Program/ United Republic of Tanzania, "Managing Sustainable Growth and Development in Dar es Salaam: Final Report of the Evaluation Mission." 1995.

[34] Halla, F. A "Coordinating and Participatory Approach to Managing Cities: The Case of the Sustainable Dar es Salaam Project in Tanzania." *Habitat International* No.18, Vol.30, 1994: 19-31.

[35] Nnkya, T. J., "Planning Practice and Democracy in Tanzania." Unpublished *Ph.D.* Thesis, Institute of Town and Landscape Planning - Copenhagen Denmark, 1996: 115.

[36] The British formulated the current urban planning standards in 1950s. Attempts to revise such standards started way back in 1960. In 1997 the Ministry of Lands prepared a draft of planning standards which are yet to be discussed and approved ready for use.

[37] Nnkya, T. J. 1996. *op. cit.* p. 120 -122

[38] Nnkya, T. J. 1996 *op cit.* p. 126-127

[39] Nkya, T. *Ibid.* p. 129

[40] United Republic of Tanzania, National Census: Preliminary Report, 1988

[41] Lerise, F. 'The Management of Rural Service Centers in Tanzania: A Case Study of Njombe Rural District" UNCHS- Habitat, Nairobi, 1993.

[42] Lerise, F., "Privatisation of Urban Services and Decentralisation in Tanzania." Unpublished report for UNCHS, *Habitat*, Nairobi, July 1998.

[43] Kironde, J. M. 1994, *Op. Cit.*

[44] Kombe, J., *Formal and Informal Land Management in Tanzania- The case study of Dar es Salaam City* Published Ph.D. dissertation, University of Dortmund, Dortmund, 1995.

[45] Lugalla, J. *Crisis, Urbanization and Urban Poverty in Tanzania, A Study of Urban Poverty and Survival Politics.* University Press of America, Boston, 1995.

[46] Stren R. and Rodney White, (eds.) *African Cities in Crisis.* Boulder, Colorado: Westview, 1989.

6

MANAGING URBAN DEVELOPMENT IN TANZANIA: ISSUES, INITIATIVES AND PRIORITIES

Fred S. Lerise and Suleiman Ngware***

BACKGROUND

This chapter provides an inventory of trends, initiatives and priorities in urban-focused development in Tanzania, with a view to enabling policy makers, researchers, bilateral development promoters and other people interested in Tanzania's struggle for development to critically assess their role in managing urban development in Tanzania. The chapter is based on a study carried out by three consultants from the Institute of Development Studies (IDS) University of Dar es Salaam, the Department of Urban and Rural Planning at the University College of Lands and Architectural Studies, and from the Swedish Embassy. Conclusions made in this chapter are derived from meetings and interviews with different people from institutions interested in urban planning and development in Tanzania.

Urbanisation is an issue in Tanzania not only because of its social and economic significance but also because of its high rate and lack of effective mechanism to manage its outcomes, such as the pressure on urban services and infrastructure. From 1980 to 1997, the urban population grew at a rate of 10% per annum. Both the 1978 and 1988 national census issues show that a bigger share of urbanisation is a result of rural-urban migration (URT, 1998).[1] It is anticipated that 17.5 million people or about 36.3% of the total population in Tanzania will be living in urban areas by the year 2001. By and large, urbanisation has become part of the development landscape in Tanzania where hundreds of young men and

* Senior Lecturer, University College of Lands and Architectural Studies (UCLAS), UDSM
** Associate Professor, Institute of Development Studies, University of Dar es Salaam.

women move from villages to urban centres every year. Dar es Salaam alone is said to receive between 100,000 to 300,000 people a year. Most of these migrants live in un-serviced parts of towns with environmental conditions which threaten their own life and health. A substantial number of the growing urban population fail to obtain their daily needs. After the implementation of the structural adjustment programmes, liberalisation and the removal of subsidised services in urban areas, poverty is as pronounced in towns as it is in villages. Attitudes and policies which perceived urbanisation as an unwanted phenomenon have changed towards accepting urbanisation as an inevitable process that provides not only challenges but also opportunities to central and local governments, bilateral and multilateral donors, the local communities as well as individuals who, in one way or the other, take part in the process.

Because of the importance and the urgency of charting out strategies to manage the fast growing urban population, a decision was made by Sida to initiate a programme to develop a multi-sectoral approach towards assistance to urban development. Similarly, the Embassy of Sweden has identified urbanisation and urban issues as one of its priority areas.

A process to enhance Sida's involvement in, and assistance to, the urban sector and urban issues in Tanzania was thus initiated in 1998. The first activity was to organise an inventory workshop in Dar es Salaam, in which knowledge and experience concerning urbanisation and urban planning and management was gathered. In addition, the workshop provided an opportunity to the Embassy of Sweden to establish contacts and networks with other institutions and individuals with interest in the process of urbanisation and its management in Tanzania. Papers presented and discussed in the workshop form a larger part of this book in the hope that they provide a knowledge base on urbanisation and its related socio-economic and management issues to the different actors and interested parties. The second activity was the mapping of urban assistance, competence and priorities among major urban planning and development partners to Tanzania. This chapter therefore presents findings from that mapping.

CONCEPTUAL AND METHODOLOGICAL FRAMEWORK

As indicated earlier, the purpose of this study was to produce an inventory of current and possible future trends, vision and priorities concerning urban-focused development collaboration activities in Tanzania. Initially, the consultants prepared a work plan and a provisional list of institutions to be visited. However, it must be realised that time for carrying out the task was rather short, thus making it difficult for some institutions and individuals, particularly international agencies, to reschedule their commitments within the duration of the study. Nevertheless, we hope that the issues raised and suggestions made in this chapter will shed some light into areas or aspects of urbanisation and urban planning which the government of Tanzania and its development partners may wish to put more emphasis on improving their capacity as well as that of community based institutions in managing the process of urbanisation in Tanzania. It is also anticipated that the findings will facilitate identification of potential pilot projects in selected towns. Lessons from the pilot projects will be used to formulate a long term programme.

ISSUES, INITIATIVES AND PRIORITIES IN MANAGING URBANISATION AND URBAN PLANNING IN TANZANIA

This section presents the issues and priority areas for possible collaboration between Sida and Tanzania, all based on the aforementioned study. The existing hierarchy of settlements in Tanzania is used to structure the issues. The presentation starts with the national level urbanisation issues because they cut across different urban settlements and sectors. At the top of the hierarchy is Dar es salaam City, which is the only urban centre with well over two million inhabitants as per the 1988 census. The remaining 19 regional headquarter towns occupy the second and third position in the hierarchy; nine of which have acquired municipality status and thus occupy the second level in the hierarchy of settlements in Tanzania. Ten regional headquarter towns have a town council status and are regarded as the third category.

Compared to the city and the municipalities, the town councils and townships have received the least attention in interventions geared towards better management of urbanisation and urban planning. In general, there has been very limited intervention in urban planning in the third category of settlements. The chapter discusses these issues in detail by first focusing at the central government.

Administration of Urban Centres

In principle, all the ministries play a role in managing urban development in one way or the other. However, the Ministry of Regional Administration and Local Government (MRA&LG) and the Ministry of Lands and Human Settlements Development (MLHSD) play a key role in influencing the administration of settlements and in carrying out urban planning functions respectively. The civil service reform and particularly the local government reform is coordinated by the newly established Ministry of Regional Administration and Local Government while reforms in land administration and in urban planning are guided by the Ministry of Lands and Human Settlements Development. We begin with the local government reform based on issues raised in interviews with officials responsible for the local government reform programme.

Improving the Administration of Urban Settlements

The government regards urbanisation and urban development as an important process for the development of the country as a whole and has decided to pursue strategies aimed at improving the management of urban centres. Thus, several pieces of legislation, including Acts nos. 7, 8, 9, 10 and 11 of 1982 providing for the establishment, organisation, administration, and financing of urban and district councils are being reviewed to reflect, among other things, the changing government attitude and priorities towards urban development. Simultaneously, the Ministry for Regional Administration and Local Government is in the process of restructuring itself. A new department dealing with urban planning and urban development has already been established in the Ministry. At the local authority level, an additional standing committee responsible for urban development issues will be created not only in urban councils but also in district councils to deal with the management of fast growing small centres. The issue is that the central government recognises the role of the small trading centres in the process of urbanisation and socio-economic development and therefore it is re-organising its institutional set up to become more effective in managing such centres. Before such committees were established the administrative status of small towns within rural districts was not very clear.

Gradual Decentralization of Service Provision

Within the decentralisation process, which is also part of the civil service and local government reform process, governance of urban settlements particularly the management and provision of municipal services will be gradually hived off from central government institutions to urban and district councils. According to officials in the Ministry of Regional Administration and Local Government, 35 district councils have been identified for a pilot project in which staff working in the health and education departments were to be fully decentralised to the councils by 1999. Thus, once that process is completed, the provision of health and education services will entirely become the responsibility of the respective local authorities. Since provision of municipal services such as housing, sanitation, education, health and others is a controversial issue facing urban planning (Kironde, 1998),[2] decentralisation efforts by the central government shall be rated successful depending on the extent to which local authorities are able to take up that responsibility. Otherwise decentralisation will be a failure if the number of urban inhabitants living in poorly serviced areas continues to increase.

The then Permanent Secretary in the Ministry of Regional Administration and Local Government seemed to be aware of the importance of decentralised service provision and cautioned that the most important issues which her ministry has to deal with were:

a) Widening the revenue base for local authorities, so that they obtain sufficient funds to meet the costs of financing their new responsibilities and tasks;
b) Related to improved financial capacity is the question of human resource capacity. Currently district councils lack sufficient numbers of well trained staff. The shortage is much more severe in the land and urban planning departments. Among the challenges facing the Local Government Services Commission, the national institution responsible for recruiting and training personnel as well as preparing incentive schemes for local government employees, is that of improving the human resources capacity within local authorities;
c) Within the local government restructuring programme, the Ministry of Regional Administration and Local Government realises and would like to emphasise that each council is unique and thus has specific needs and potentials. The practice of pan-territorial local government structure where each council is supposed to be organised into twelve technical departments and six standing committees will be reviewed. The ministry sees the need for a new organisational structure of local authorities based on the individual contexts of such authorities.

Decentralized Urban Planning

From interviews with officials in the Ministry of Lands and Human Settlements Development, it is evident that there is a clear change of attitude towards more decentralisation of urban planning activities to local authorities than before. In general there appears to be an agreement that both long-term broad strategic urban development planning and short term detailed planning at local level should be carried out by respective local authorities. The role of the ministry is essentially that of providing policy framework and when necessary carrying out activities which cannot be undertaken by individual local authorities. Following that change, a number of strategic changes have been made: The Human Settlements Development Department in the Ministry of Lands and Human Settlements Development has established a section whose main task is to assist local authorities to formulate long term strategic development plans for their urban areas. In some instances, the ministry staff work hand in hand with staff in the local authorities. For example, at the time of doing field work for this chapter, two planners from the department were working with the Tabora municipality in the preparation of an urban strategic development plan for the town. Thus, preparation of urban plans is no longer the sole responsibility of the ministry or private consultants alone.

To really effect the decentralisation of urban planning, relevant pieces of legislation need to be revised. Planners and other related staff have to be deployed to local authorities especially in district councils; but the question of how to deploy competent staff at district level remains unresolved. At the same time, if what the the Acting Director for Human Settlements Development Department observed is correct, namely that the majority of staff in the regional offices will shift to the district level and that some staff may actually shift to districts of their choice, there is the added problem of districts in remote areas not being able to attract experienced staff from the regions.

In addition to the above drawback, some local authorities do not like to recruit staff from their respective regions, for a number of reasons not least among them being the fear of tribalism. Some local authorities are warry of employing regional staff who are close to retirement age. In order to improve the human resources capacity in the local authorities, the following issues have to be dealt with.

Lack of Sufficient Number of Qualified Staff

The quality of manpower in the departments responsible for lands and urban management is rather low. Most officers who head the lands development department at district level - land valuation, town planning, and land surveying - are certificate holders trained a long time ago. Although the University College of Lands and Architectural Studies (UCLAS) produces graduates in these fields, few district councils have employed such staff. For instance, none of the four rural districts in Lindi and three in Mtwara regions has employed a town planner. A massive employment programme may have to be considered in order to improve the staffing position in the district councils. This massive employment should however be done cautiously if only to avoid the common practice of fresh graduates with little or no work experience being given the task of opening and running a new district town planning office.

Lack of Equipment and Office Facilities

In addition to dealing with the manpower issue, the central government and local authorities have to seriously consider the question of office facilities and working equipment. One of the officials gave an example of land registration and issuance of title deeds as one of the tasks to be decentralised. But he was concerned that, although that is a simple task, most local authorities lack equipment to carry out such an important task in land use planning and land administration. According to the official, mapping is central for land registration and land use planning. Effective urban planning largely depends on availability of accurate and up to date maps. It will thus take time before local authorities are able to issue title deeds, unless there are real efforts to improve facilities needed by local authorities.

In an attempt to address the questions raised above with respect to manpower, facilities and equipment, the Acting Director in the Department of Human Settlements Development in the Ministry of Lands listed the following priority activities and potential areas of collaboration between external development partners such as Sida and the Tanzanian government through the Ministry of Lands:

a) Organisation of seminars, workshops, training sessions etc. in order to inform local authorities, communities, and individuals about the process of decentralising urban planning, and make them aware of their new responsibilities in managing urbanisation and in carrying out urban planning. People and their local institutions should be well informed that they are the controllers of their own development process in their areas and that they should know how to plan and mobilise resources to realise their plans;

b) Review of the 1956 Town and Country Planning Ordinance (as revised in 1961) and other relevant pieces of legislation (for instance building regulations), to reflect social, political, economic and physical changes in the urban environment. The proposed review should also take into account provisions stipulated in the new Land Act, (enacted by Parliament in early 1999). The Human Settlements Development Policy will also form an input in revising the Ordinance. There is general concesus among policy analyists, practitioners, and observers of Tanzania's reform process that the new urban planning legislation should have a wider scope than the existing one, particularly in incorporating human settlement management issues into the planning laws;

c) Another priority issue with respect to urbanisation and urban planning is the question of affordable and decent shelter, especially for the urban poor. The Human Settlements Department puts its priority in interpreting the new Land Law in terms of its provisions which can be used to support initiatives aimed at improving shelter provision in fast growing urban centres. A number of people showed concern at the Parastatal Pension Fund, which invests pensioners' funds without taking into account the interests of the main contributors, for instance in enabling contributors to have affordable shelter when they retire. Housing cooperatives was also mentioned as a possible strategy through which low income households can have access to better and affordable shelter;

d) In order to build staff capacity in the districts, the Ministry of Lands and Human Settlements Development has prioritised restructuring the organisational set-up of district councils so that the technical and management staff can work more closely in developing and implementing urban development plans.

Urban Planning Issues in the City of Dar es Salaam.

Dar es Salaam City, which is made up of three administrative districts and accommodates a population of about three million people is the *de-facto* capital of Tanzania. In addition to the re-introduced City Council which had been in abayance since 1996 following its abolition by the central government, the city is also administered by a Regional Commissioner. Following proposals made within the local government reform programme, the City Council has been restructured into three municipalities in line with the existing administrative districts. Once the restructuring exercise is accomplished, urban planning activities within the city will be shared among the three municipalities and the City Council.

Before these reforms, the erswhile City Commission was responsible for both long-term broad urban planning as well as short term detailed local planning. In carrying out its urban planning functions, the Commission was supposed to be guided by the 1979 Dar es Salaam Master Plan. Efforts by the Ministry of Lands and Human Settlements Development to revise the 1979 Master Plan led to the establishment of the Sustainable Dar es Salaam Project in 1993.

Experience from the SDP subsquently led to the creation of an Urban Authorities Support Unit (UASU) in the Ministry of Regional Administration and Local Government. The unit is responsible for assisting urban authorities to reform urban planning. At the time of writing this chapter UASU was working with the City Commission to prepare a strategic development plan for the city. The plan will thus replace the 1979 Master Plan which, in many ways is out of date. However, despite efforts by the central government, UNDP, HABITAT, the World Bank, JICA, Irish Aid, and many others to deal with the outcomes of urbanisation in Dar es Salaam, very little has been achieved in terms of poverty alleviation, improved access to municipal services, and better living conditions.

For the past ten years or so, over 70 per cent of the city residents are accommodated in high density areas with room occupancy rate as high as eight people. Lack of reliable water supply, sanitation, drainage, security and other municipal services threaten public safety and health in urban settlements. Problems associated with poorly planned and uncontrolled spatial growth of the Dar es Salaam City are likely to increase if the city continues to grow at the present rate of 8 to 11 per cent per year. For instance, for the past three years especially after trade liberalisation and the retrenchment programmes, the city receives about 300, 000 people annually. In other words, population wise one big municipality (say Mbeya or Dodoma) is added to Dar es Salaam every year. Thus the issue facing the City Council and its collaborating partners is how to manage such a big increase in urban population and how to chart out strategies to enhance and utilise potentials created by the big agglomeration of people who are not only able to work but also provide a big market for urban services.

The Director of Human Settlements Development, Professor Hayuma, associates the low achievements partly to the uniqueness of Dar es Salaam in terms of high population growth, political as well as economic powers, and partly due to lack of knowledge about the development dynamics of the city. He further argues that, researchers have not sufficiently directed and focused their research towards findings which can be directly used by decision makers in managing urbanisation and urban planning. That type of knowledge is critically required but does not exist. Thus research to feed in urban planning practice is also a priority area.

In addition to the lack of a city level long-term plan (which will hopefully be taken care of once the UASU accomplishes the preparation of the Strategic Development Plan), the following priority issues were raised in interviews with officials of the then City Commission:

a) The majority of the city residents are accommodated in poor housing conditions. The Commissioners associate that situation with the absence of big real estate developers who can carry out large scale housing development and build affordable housing for the city residents;
b) Upgrading of squatter settlements to improve the spatial organisation and improve the availability of services to residents;
c) Improve the overall sewerage system in the city;
d) Increase the involvement of the private sector and community-based organisations in municipal service delivery. The erstwhile Commission had started with sewerage, garbage collection, and city cleaning;
e) Carry out re-development projects in old parts of the city in order to improve roads and street systems and therefore attract large scale property development in the city centre, for example in Kariakoo.

When asked about specific areas of collaboration, the Commissioner for Urban Planning was of the opinion that there is ample room for external collaboration and suggested that a thorough study should be carried out in order to identify the exact situation. On the basis of that study actual areas and projects for collaboration could be identified.

Urban Planning Issues in Municipalities

Mainland Tanzania is divided into twenty administrative regions. Each region has a regional headquarters town. The twenty towns experience different rates of population growth and thus urban development. Excluding Dar es Salaam, nine such towns are classified by the government as municipalities. Probably because of their status in the hierarchy of settlements, the nine municipalities are given more attention in urban planning than their ten counterparts. The municipalities are geographically distributed as follows: In the south are Mbeya, and Iringa. Tabora, Dodoma and Morogoro are in the central part, while Mwanza is in the Lake Victoria zone. Arusha, Moshi and Tanga are in the northern zone.

Population and the growth rates for these municipalities vary. According to the 1988 national census, only Mwanza and Dodoma recorded a population of over 200,000 people. Ten years after the census, i.e. in 1998 the population of Mwanza municipality was estimated at 420,000 people with an annual growth rate of about eight per cent. Information from the Mwanza Environmental Profile (1998)[3] suggests that most of the population growth is due to rural-urban migration. Municipalities with population between 100,000 to 200,000 in the 1988 census are, Arusha, Mbeya, Tanga and Morogoro. The rest namely: Moshi, Tabora and Iringa had population ranging from 84,000 to 96,000 people as per the 1988 census.

Urbanisation and urban planning issues in these municipalities are not significantly different from those experienced by residents in the city of Dar es Salaam. The only difference is the magnitude or intensity of such problems. Unlike Dar es Salaam, the municipalities are managed by elected councils made up of councilors elected from each administrative ward and the locally elected Members of Parliament. The Councilors elect a Mayor, as their leader. The council operates through a committee system. Following the provisions of the Local Government (Urban Authorities) Act, No 8 of 1982 - which is currently under review - the statutory committees found in each municipality are: (1) health; (2) trade and economic services; (3) urban planning; (4) human resources deployment(5) public works; (6) education and culture; and (7) finance and administration.

In managing the municipalities, councilors are supported by a municipal director who is appointed by the Ministry of Regional Administration and Local Government. The main responsibility of the director is to supervise and coordinate activities of other departments. Most municipalities have nine technical departments:(1) health and social services; (2) cooperatives; (3) trade, commerce and economic affairs; (4) urban planning; (5) community development; (6) public roads and buildings; (7) education and culture; (8) finance and personnel; and (9) administration.

As in Dar es Salaam, urban planning issues in the nine municipalities are supposed to be guided by urban master plans. Master plans for the nine municipalities were prepared in 1974 by the government in collaboration with donors. A large share of the support was provided by Finland. Some of the plans, for instance those of Arusha, Tanga, Mwanza, Morogoro and Mbeya were revised in the early 1990s. Efforts by the Ministry of Lands to review the Tabora and Moshi master plans are underway. In addition to such urban planning initiatives from the Ministry of Lands, the Urban Authority Support Unit, is also assisting these municipalities in developing a more participatory urban planning approach as well as in encouraging the formation and involvement of community based organisations.

In order to improve the performance of urban planning, the new participatory approach works towards enhancing grassroots involvement through participatory tools to ensure real consultation and that different needs and local potentials are identified and utilised. In Iringa, for example, in addition to municipal level consultations, DANIDA is considering carrying out ward level consultations in order to bring the urban planning process closer to the residents. Workshops and publication of urban plans and other related documents; as well as public presentations in Kiswahili, are likely to make more urban residents interested and thus participate in the planning process.

The UASU staff are aware of the important role played by the Ministry of Lands in reforming urban planning practice in Tanzania at policy level. They therefore have decided to involve town planners from the Human Settlements Department in the Ministry of Lands in their programmes as much as feasible.

Discussions with key UASU staff suggest that the unit intends to assist municipalities not only in reforming urban planning and the management of urban development processes, but also in securing funds to implement municipal plans. For instance, UASU has assisted Mwanza and Iringa municipalities to obtain financial support from DANIDA. The unit has also assisted Dar es Salaam City to obtain financial support from the World Bank to finance community infrastructure improvement projects in the poorly serviced areas of the city. Since UASU works with different donors, a major area of concern must be that of coordination between donors, government institutions, municipalities, communities and the unit itself. But according to the coordinator of the unit, there are no coordination problems and if and when they arise, meetings to discuss the issue and iron out the differences would be held.

The Counsellor for Development in the Royal Danish Embassy was of a different opinion, however. He admitted that DANIDA has faced some problems with UNDP in the process of supporting Mwanza and Iringa municipalities. According to the Counsellor, there is a need for clear guidelines on the modalities of cooperation between UNDP and DANIDA. Within such guidelines, modalities of cooperation should indicate for instance, when does a municipal council approach the UNDP and when does one go to DANIDA or any other donor? The Counsellor emphasised that before a donor provides support and interventions to a municipal authority, a feasibility study should be carried out. At the same time it was insisted that double or overlapping donor support to a municipality should be discouraged and that funds in a support programme should be controlled or managed by a single donor.

The coordination issue raised by DANIDA should be a concern for other donors not only because DANIDA has already acquired experience in running collaborative projects in the urban sector, but also because there are many donor funded projects in the municipalities and in other urban areas, thus necessitating an effective coordination mechanism between the different actors. In Mwanza municipality for instance, there are at least eight donor supported projects related to the environment and urban planning namely: capacity building in environmental management by DANIDA; urban sector rehabilitation programme by the World Bank; Lake Victoria Environmental Management project by the Global Environmental Facility (GEF) and the World Bank; institutional strengthening for urban water supply and sewerage improvement by the European Union; health, sanitation and water under UNICEF; urban planning and service provision activities carried out by Plan International; and similar initiatives by the Local Initiative Facility for Urban Environment.

The now familiar situation of many donors working in one municipality, puts pressure on the government in terms of coordination. That task is enormous for the UASU which has hardly worked for two years. If the unit is to become more effective in such an important and challenging task, its capacity has to be substantially enhanced. When asked about the main areas of possible collaboration with donors, the unit staff listed the following in their order of prority:

a) Equipment (digitizers and computers) to facilitate mapping at municipal level, because as already pointed out elsewhere, mapping and land registration is a real problem facing municipal councils in land administration and in land use planning;
b) Training of municipal staff;
c) Support to municipalities like that provided to Iringa and Mwanza by DANIDA.

Urban Planning Issues in Regional Towns and Town Councils

This section focuses on towns with a town council status. These include the ten regional headquarters towns and a few district councils, such as Kilosa and Korogwe which have been upgraded into town council status. Regional headquarters towns which fall into that level are Shinyanga, Sumbawanga, Songea, Kigoma and Singida with a population of over 80,000 people, and Mtwara, Lindi, Musoma, Bukoba and Kibaha with a population ranging from 37,638 (Kibaha) to 76,632 (Mtwara) as per the 1988 census.

This level of urban centres has not received much attention in terms of urban planning interventions by the government or donors. With the exception of the preparation of master plans or interim land use plans by the Ministry of Lands or by local consultants upon commissioning by the Ministry, there has not been other urban planning efforts by the central government directed to these towns. Most of the plans, if available, are either out of date or only indicate land sub-divisions and thus are not really useful for guiding urban planning discussions in those towns. Although a considerable number of donors operate at district level, their interest has been sectoral, mainly focusing on natural resources management, agricultural production and rural development. To be sure, some local authorities in this category are keen to prioritise the effective management of urban areas in their districts, but donor interests and agenda seem to override those of the local authorities because of their financial muscle and the later's dependency on them. Among the outcomes of this neglect of the towns is that urbanisation is hardly managed and the capacity of the local authorities to deal with urban planning is equally underdeveloped. However, the proposal to establish urban planning committees in local authorities with fast growing urban centres may improve the management structure of the town and district councils. The issue of manpower and human resources capacity in these local authorities has to be addressed if these committees are to be effective.

Urban Planning Issues in Townships and Trading Centres

Below the town councils are other smaller urban centres. Some have acquired township status, others have not. Whatever the situation, the small towns are performing important urban functions to their respective hinterlands or villages. The number of these centres is not known but they include about 90 district headquarter towns, about 128 small urban centres as listed by the draft National Human Settlements Development Policy. Mascarenhas (1998)[4] questions whether there are only 435 trading centres in Tanzania. There are also many villages which are performing some urban functions.

In general this category of settlements hierarchy `accommodates the bulk of urban activities but they are the least in terms of urban planning interventions. They not only lack urban plans but also lack local level institutions responsible for urban management issues in those towns. There is no proper governance and it is not clear as to which authority is in-charge in such centres. In most districts, for instance Kibaya, the urban area is administratively regarded as one of the wards and thus represented in the district council by a ward councilor who may not be able to provide sufficient attention and interventions for the center. Hayuma of the Ministry of Lands listed Njombe, Mbinga, Tarime, Bunda, Nzega, Ifakara and Babati as some of the fast growing urban centres in Tanzania, which neither the government nor donors have assisted in terms of urban planning. Most of the people are of the opinion that more efforts should be made to improve the management of the small towns as one way of decentralising urbanisation and urban planning.

RECOMMENDATIONS

Urban and Rural Areas as Complementary

The interview responses support Sida's and other donors' intention to assist the government and communities in managing Tanzania's urbanization and urban planning. Although there are many donors collaborating with the government and other actors in the urban sector, given the magnitude of the urbanization-related problems and Tanzania's large size there is a big room for more donor-based collaboration in the sector. However, for donor support to be effective the philosophy and approach should regard rural and urban areas as two interdependent settlements. They have different requirements but depend on each other. The approach has to combine rural and urban initiatives in order to be more effective in managing urbanisation and its outcomes.

Encourage Different Experts to Work Together

As repeatedly emphasised in this chapter, the key issue in the management of urbanisation and in improving urban planning within a decentralised system, is to build the capacity of local authorities as well as that of the central government institutions to support the local authorities carry out the reforms. In this context the consultants are convinced that for Sida support to be effective, capacity building in the central government and in local authority levels should be the main focus of the collaboration efforts. In relation to capacity building, one of the guiding questions in charting out the form of collaboration is: how to organise the administration structure of local authority departments into broad management teams which will encourage (or force) the different experts to work together; and how the staffing should be in order to deal with urbanisation and urban planning issues. In other words, how to bring the reform process to the local level.

Comprehensive Urban and Rural Planning Project

Sida has substantial experience in supporting the management of urbanisation and in improving urban planning from her involvement in urban planning initiatives in Kimberly and Port Elizabeth in South Africa. However, since towns differ a lot, pilot projects to develop experiences relevant to many Tanzanian towns might be a good idea to start with. It is thus recommend that a pilot project in a comprehensive participatory urban and rural planning exercise in a district be undertaken. Through such pilot project, Sida could collaborate with the local authority, the central government, local communities found in the area or established in the course of the implementing the project. Sida can also collaborate with other donors in not only improving the capacity of the different actors but also in producing long term plans as well as local detailed plans and in securing local as well as external resources to implement the plans.

In discussions between the authors of this chapter and the head of the Urban and Rural Planning Department at UCLAS, Bagamoyo District located about sixty kms north of Dar es Salaam city, was mentioned as among the district councils which have approached the department for support in capacity building in urban planning. In addition to the request, the town is facing pressure from local as well as international developers and thus there is a need to put in place, sustainable ways of managing that development. To complement the lessons from Bagamoyo another district with different urbanisation and urban planning issues from those of Bagamoyo should be included in the pilot project. Kibaya (Kiteto District in Arusha), where Sida has been active for a number of years, is a potential candidate for consideration.

Research and Training as Part of the Pilot Project

In order to facilitate effective dissemination of lessons learnt from the pilot districts, the collaboration venture should include a research and training component. Through workshops and seminars, short and long term courses, human resources capacity will be improved gradually. Networking with other districts and similar projects in other countries could also form a source of additional knowledge needed in the reform process. The already existing networks and collaborative activities between Tanzanian research and training institutions and similar institutions in Sweden could be utilized and enhanced to provide much needed inputs into the pilot project and to the dissemination efforts. Through research and training, the project can also offer assistance in reviewing urban planning related legislation in Tanzania.

END NOTES

1 United Republic of Tanzania, "Draft National Human Settlements Development Policy." Dar es Salaam, Ministry of Lands, May 1998

2 Lussuga Kironde, *An Evaluation of Government's Attempts to Cope with the Growing Demands imposed by Rapid Urbanization in Tanzania*, a paper presented at the workshop to establish a knowledge base on urbanization in Tanzania - September 2-3, 1998 Dar Es Salaam.

3 Mwanza Municipal Council, DANIDA Kampsax International A/s, *Environmental Profile of Mwanza Municipality* June 1998.

4 Mascarenhas, A. " Understanding Historical and Current Causes for Rural-Urban Migration in Tanzania." Draft working document foı a workshop to establish a knowledge base on urbanization in Tanzania - September 2-3, August, 1998.

7

PROMOTING GOOD URBAN GOVERNANCE AT COMMUNITY LEVEL: THE CASE OF TABATA, DAR ES SALAAM

*Joseph D. Kessy**

INTRODUCTION

Tabata is a suburb situated to the South Western fringe of the city of Dar es Salaam approximately 8 kilometres from the Central Business District. The area used to be referred to as "Kisiwa cha Wavumilivu" whose literal translation means island of the patient people. With a population of over 60,000 people Tabata is famous because the main site for dumping waste collected within the jurisdiction of Dar es Salaam City Council used to be located at Tabata along the Nelson Mandela express way.

Tabata is a newly planned residential area accommodating thousands of people. Some parastatal organisations, namely the National Social Security Fund, National Insurance Corporation and Tanzania Zambia Railway Authority have erected substantial housing estates inthe area. This is because houses are owned by individuals and some government departments. The value of the land within Tabata community used to be low because of the presence of the dump - a user activity which is not compatible with residential use.

COMMUNITY REACTION

The Tabata community is vehemently protective of its right to a good environment; and for this reason they fought as a 'pressure group' to ensure the refuse dump at Tabata was closed. The community fought through the Courts of law because continued use of the dump site at

* Property Valuation Consultant and Chairman of the Tabata Development Fund (TDF)

Tabata posed serious health hazards to the community - particularly the pungent smell from polluted air and toxic substances emanating from burnt wastes/rubbish. Motorists driving along Nelson Mandela expressway and pedestrians were also exposed to these hazards because of the smoke coming from the dump site often also blocking their vision. In addition, food stuffs unfit for human consumption were always dumped at the site, but later on scavengers used to re-introduce them into the city's food chain, thus endangering the life of City consumers. With environmental preservation awareness, the Tabata community collectively fought to ensure the dump site is removed from the area. On 31st August 1991 the High Court ordered the closure of the Tabata dump site.

Taking into account the fact that Tabata community was fighting as a pressure group, we decided to form a society with legal status known as the Tabata Development Fund (TDF) - a community based organisation. The CBO was subsequently registered in March 1993 by the Registrar of Societies, Ministry of Home Affairs. Therefore, good governance within our community started when we formed the society because the need to have a registered society (CBO) was not imposed onto the community by any body. The organisation operates using its constitution and currently there is a TDF executive committee, six neighbourhood representatives and over 150 area representatives each representing more than 10 households (in surveyed areas) and over 15 households (in squatter areas). In terms of gender the organisation is not gender blind; rather it observes gender balance insisting on equal representation.

OBJECTIVE OF TDF

The main objective of TDF is to upgrade deficient infrastructure within Tabata through self help. Translation of this objective into action has so far resulted into the rehabilitation of 4 kilometres of the Tabata-Segerea road and planting of trees in an attempt to preserve the environment. In order to reinforce security in the area there is a committee dealing with peace, in connection with which about 200 torches and dry cells were at one time provided to night security guards (*sungusungu*). There are seven (7) more sub-committees under TDF dealing with education, social welfare, technical, finance, health and environment, and trade and economic affairs. The good performance and governance at community level led to international recognition through Community Infrastructure Project (CIP) - an offspring of the Sustainable Dar es Salaam Project (SDP). TDF is now a partner in development at community level with Dar es Salaam City Commission, the Irish Aid, the World Bank, and utility agencies like TANESCO, TTCL, and DAWASA.

PROJECTS UNDERWAY

It is the community members who decide on the type of project to embark upon depending on the need and resources available. Going by the 'preference scale' the Tabata community decided to tackle the problem of water, and roads. DAWASA water cannot suffice. As a result we have embarked on digging two boreholes and establishing a water network system to reach the community by providing water pipes at various domestic points. The first borehole has a capacity to produce about 3,000 litres of safe domestic water (clean water) per hour and it serves about 4,000 people in neighbourhood number 5 - a squatter area with the most deficient infrastructure in the area. There is a reserve tank with a capacity to store about 30,000 litres of water. In order to maintain the water pipes and pay the stand pipe attendants we charge T.shs 20 per 18 litre bucket of water. We are getting about T.shs. 300,000/= per month from this endeavour.

The second borehole, with a capacity to produce about 7,000 litres of clean water per hour is in the final stages of completion. It has a storage capacity of about 60,000 litres to serve over 10,000 people. TDF expects to realize T.shs. 600,000/= per month from water sales from this project. The capital cost of construction for first borehole is Tshs. 15 million and that of the second borehole is Tshs. 27 million. Irish Aid has provided funds for purchase of building materials while labour (digging of trenches etc.) and supervision is provided by the community through self help. As part of consultative/ participatory management it is the community which decided where to locate the boreholes, the water network system, and how much should be charged for each 18 litre bucket of water.

There is also a 3.3 kms tarmac road project for the spine road from Tabata to Segerea, and 16 kms of neighbourhood gravel roads. We have signed a memorandum of understanding with the Dar es Salaam City Commission in which we are 100% responsible for maintenance and operation of the neighbourhood roads. Furthermore, the community is to raise Tshs. 38.8 million being the 5% counter-part fund of the capital cost of constructing the neighbourhood roads. Each household contributes Tshs. 17,000.00 towards the capital cost of construction of the neighbourhood roads. The remaining 95% is the counterpart fund from the World Bank. Other partners to the roads project are the Dar es Salaam City Commission, Irish Aid, and other utility agencies like TANESCO, TTCL etc.

PROBLEMS

Our community has people of different cultures, religions, tribes and different levels of education, purchasing power and income. Not all community members participate in the development activities. For instance the Tshs. 17,000/= (5%) contribution per household for the construction of the neighbourhood roads, there are some community members who are arguing it is the government's or Dar es Salaam City Commission's responsibility to rehabilitate the roads and not the community. Others are arguing that they don't have vehicles therefore people with cars are supposed to contribute more. Above all there are people who are willing to contribute but they don't have the economic power to contribute effectively. We are trying our best to sensitise people through seminars, meetings and leaflets on the need to contribute and we are happy to say that more than 80% of the community members are satisfied and willing to pay and that more than 50% have contributed.

SUCCESS

Tabata community members are active and not passive in matters concerning their own development. We define our needs and we do what we have chosen and the government and the City Commission do not impose on us what should be done. People are informed and aware of the need to involve themselves in deciding what is to be done at community level through self help. There is a sense of ownership and confidence by the community members to the CBO and its leadership because whatever is done there is transparency and equal representation in decision making. Above all else, Dar es Salaam City Commission does not impose any decisions concerning our own development, rather it is our partner in development.

Indeed, the Tabata community, under the coordination of TDF, KIJIKO and CIP has enabled the Dar es Salaam City Commission to win the **1998 Dubai International Award on Best Practices** which took place in Dubai during the World Habitat day. The Dar es Salaam City Commission was represented by the coordinator - CIP, and the Chairman Tabata Development Fund during the awarding of a trophy, a certificate, and cheque amounting to US Dollars 30,000 out of 10 municipalities which won the award world wide.

8

RAPID URBANISATION IN THE 21ST CENTURY: FUTURE RESEARCH ISSUES

*Rose Shayo**

INTRODUCTION

Available projections suggest that by the year 2010 the majority of the world population will be living in urban centres save for most low income countries in the Third World (Stren, 1994; Swirling, 1997). Swirling (1997) for instance, argues that although two thirds of the population in the developing countries, particularly Sub-Saharan Africa, will be living in rural areas the percentage of urban population will be increasing. In East Africa, the urban population is expected to increase from 29 % in 2,000 to 36.0% by the year 2010, that is an increase of about 7 percent. Interestingly, this rate is more or less similar to other regions in the world (see Table 8.1).

Discussions on the nature and quality of urbanization are gaining momentum throughout the world, including Tanzania. The key question being posed by many people is: "Are cities going to be saviors, villains, or by-standers? (White, 1996). More precisely, many people are concerned with the rapid urbanization taking place in an environment of increased urban economic distress. In most countries, the current wave of macro-economic reforms is creating difficult conditions, both for governments and their citizens. Indeed, it is becoming increasingly difficult for the governments in Sub-Saharan Africa to create and maintain a socio-political environment climate conducive for promoting sustainable urban development.

* Lecturer, Institute of Development Studies, University of Dar es Salaam.

This chapter seeks to examine the main critical areas of concern necessary for creating an urban friendly environment in Tanzania. The main aim of the analysis is to:

- Identify critical areas of concern where policy intervention should concentrate on; and
- Propose strategies or viable options on strategic entry points.

Table: 8.1 *Percentage of population living in urban areas by world region: 1990 – 2010*

	Percentage Urban Population			Average Annual Growth Rate	
	1990	2000	2010	1985-90	2005-10
World	45.2	51.1	56.5	3.1	2.3
Western World	72.6	74.9	77.9	0.8	0.7
Developing Countries	37.1	45.1	51.8	4.5	2.9
Africa	33.9	40.7	47.7	5.0	4.2
East Africa	21.8	29.0	36.0	6.8	5.1
Central Africa	37.8	45.6	52.4	5.1	4.6
North Africa	44.6	51.2	57.7	4.0	3.1
Southern Africa	54.9	61.3	66.8	3.6	2.8
Western Africa	32.5	39.8	47.3	5.5	4.6
Latin America	71.5	76.4	79.9	2.9	1.9
Caribbean	59.5	64.8	69.2	2.6	1.7
Asia	34.4	42.7	49.7	4.5	2.6

Source: Extracted from Swirling, (1997: 14, Table 1).

At this juncture it is important to point out that the format of this chapter is rather different from the traditional one. Instead of attempting to summarise the many ideas from the different chapters presented in this book (as is always the case) I will try to extend the discussion by addressing some of the pertinent issues concerning the urbanization process in Tanzania.

The chapter is divided into three parts including the introduction. In part two we highlight the areas of concern for urbanizing Tanzania and discuss the alternative strategies for addressing the existing problems. The third part ends with concluding remarks pointing out the difficulties of locating where to begin in improving the urbanisation processes in Tanzania. Overall, having reviewed the chapters presented in this book it appears that several issues are still marginal or encompass aspects, which require more in-depth studies. However, in view of the fact that it is not possible to address all the critical issues an attempt is made to highlight the main critical areas, considered either as central, or as catalysts for the urbanization processes in Tanzania.

URBANIZING TANZANIA: CRITICAL AREAS OF CONCERN

Generally, urbanisation is considered an essential ingredient for promoting socioeconomic development. History reveals that sustained economic growth is always accompanied by the growth of urban centres. When countries develop, they promote the growth of the industrial sector, thereby creating the conditions for the growth of urban centres and other sectors such as agriculture, trade, and industry. This model is applicable in most developed countries such as France, Britain etc. but not in most Sub Saharan African (SSA) countries and Tanzania in particular. Different factors seem to account for the growth of big and small cities inTanzania. Growth of small towns such as Himo or Kibaya is quite different from others partly because of lack of big industries and poor transport network.

Both the Habitat Report (1996) and White (1996) stress that urbanization processes are known to influence socio-economic development and civilization in most countries both in the North and the South. In addition, White (1996) argues that urbanization plays an important role in promoting a sustainable environment in any setting. From this discussion, it appears that there is a positive correlation between urbanization and socio-economic development. However, several authors including the chapters by Ngware, Kironde, and Lerise in this volume, and Halfani (1994), and White (1996), have cautioned that although urbanisation appears on the face to be desirable, it is not without its own problems.

In this section, an attempt is made to focus on the critical areas of concern, which need to be given appropriate attention in order to improve the quality of urban life. The critical areas do not reflect any order of priority. As the title suggests, the entry point in prioritizing the issues is not so straightforward. This is basically due to the fact that the urbanisation crisis in Tanzania is caused by a number of multi-faceted, multi-dimensional and multi-sectoral factors that are so intertwined and linked up with both external and internal causes.

Research Based Policies and Practices on Urban Governance

Over the past decade, there has been a growing recognition of the importance of creating baseline data and information on the state of the art of big and small urban centres in Tanzania and the prospects for sustainable development. There has also been a growing recognition of the way in which urban research has been neglected in the past and its consequences on urban planning and management.

The main reason for setting the urban research agenda is to generate and accumulate useful data and knowledge that can be used for effective decisions and actions on socio-economic development. As already indicated by Ngware (Chapter 2) to date there is scanty and scattered urban research and studies in Tanzania. A central element of the current strategy for improving the quality of urban centres is easy availability of up to-date data and information on the urbanisation processes. Expanding the urban research agenda can easily do this. Based on a review of the chapters in this book, it is apparent that there is a growing complexity in the urban situation and a wide range of issues demanding concurrent research attention. However, as already mentioned in the introductory chapter, as authors we were not expected to exhaust all the research issues but rather to provide insights on the state of the art on urban research and urban studies in Tanzania.

It is important therefore to start by suggesting that the research themes, which are not covered in this book, form the agenda for future research on the urbanisation. This when combined with the existing research studies, will enable us to get a clearer understanding of the urbanisation process. Some of the critical urban issues for future research include:

- Urban crisis, urbanization and poverty;
- Comparative and interdisciplinary urban research;
- Micro level urban research focusing on community based initiatives;
- Research on retrenchment, migration and urbanization;
- Resource mobilisation for urban research; and
- Gender issues in the urbanisation processes.

Urban Crisis, Urbanisation and Poverty

It is difficult to distinguish between urban crisis, urbanization and urban poverty: The three issues are so intricately intertwined that it is difficult to separate them. For research purposes, we have tried to look at each aspect separately.

Urban Crisis

The notion of "urban crisis" involves complex meanings and interpretations and is considered to be context specific. For instance, to the government (policy makers and planners) urban crisis is reflected in the inability of the government (at national, municipal and local levels) to provide adequate services and supportive infrastructures including housing, transportation, water, shelter, schools, hospitals, and communication.

On the other hand, to the poor, the notion of 'urban crisis' is reflected in their marginality to access of resources and economic opportunities available in urban areas. In fact, for most poor people the urban crisis is manifested in the form of lack of formal/informal sector employment, inadequate access to basic facilities such as food, health, education, water, energy, transportation, recreation, housing and safe environment. Most of the internal problems cited above are increasingly becoming issues of concern for the majority of the poor people in urban areas.

In addition to these internal problems, other external factors such as structural adjustment and globalization seem to compound the urban crisis in Tanzania. More research is required to identify the spatial factors compounding the urban crisis in the era of structural adjustment, and in particular globalization. This entails a study on both the demand and supply sides of urban services, the proliferation of private investment and problems of social service delivery. This is in line with Ngware's contention that a lot remains to be done in understanding the current urban crisis in Tanzania. For example, there is need for in-depth studies which show the patterns and trends of access to social services, both from the public and the private sectors. Ngware's chapter suggests that there are very few studies, which provide useful insights on the urban crisis in Tanzania. Significantly, the existing studies focus on a narrow range of issues; such as physical planning, the environment and formal institutions. Thus, a lot remains to be done particularly on the manifestation of the current urban crisis.

Urbanisation, Poverty and the Urban Poor

In its concluding remarks, Ngware's chapter raises the question of poverty and urbanization, in Tanzania. Overall, considerable achievements have been made in improving the living conditions of the majority of the population in the Third World, including Tanzania. For instance, in most Third World countries infant mortality rates have fallen from 104 per 1000 live births in 1970-75 to 59 in 1996. At the same time life expectancy rose by four months each year since 1970. Similarly, in Tanzania, life expectancy increased from 45 years to 52 years by 1996 (GoT). There are plans to raise the life expectancy to 70 years by the year 2015.

Despite the above achievements, poverty still remains a big problem in Tanzania. With the introduction of structural adjustment programs and globalization, the quality of life of most people is going down. Available evidence suggests that most people in urban and rural areas eat one meal per day instead of three. Moreover, although the inflation rate has declined to less than 10% by early 1999, the money in circulation is very little (1999 Budget Speech). This implies that the majority of the urban poor have almost no money for meeting their daily requirements. The minimum salary has been increased from T.shs 30.000 to 45.000. The government needs to make deliberate efforts to expand and introduce effective poverty alleviation strategies that will reach and help the majority of the marginalized people both in urban and rural areas. Past efforts to address poverty problems have been criticised for being one sided in the sense that they focused more on economic welfare. Available evidence elsewhere suggests the need to broaden the scope of emphasis i.e. moving from economic achievements to social welfare in an integrated manner.

As a starting point, there is need to expand our vision by evaluating the outcomes of the poverty alleviation strategies which have been established to address the needs of the marginalized people particularly in the urban areas. It is instructive to note that the implementation of structural adjustments since the mid-1980s has significantly affected the urban population and in particular the urban poor and the 'new poor' (retrenched workers). Moreover, there is need for further research on the notion of the 'urban poor.'

The living and working conditions of most urban poor in locations such as Dar es Salaam or Moshi may be different. Thus, there is need to conduct micro level studies that compare, for example, the living conditions and access to services of the urban poor working in expatriate houses in Oysterbay, Dar es Salaam; the "machingas" and the unemployed in cities such as Dar es Salaam, Arusha and Mwanza as well as small towns such as Himo (Moshi rural), Kibaya (Arusha rural) and Kibaha (Coast Region). On the one hand the condition of domestic servants working in expatriate houses may be slightly better compared to the remaining categories. Indeed, living conditions are extremely bad for the unemployed urban people living in squatter areas in the big cities. On the other hand, the conditions of the urban poor living in similar situations in shanty towns of smaller towns like Mafuta Street in Moshi town is not known.

Comparative and Interdisciplinary Urban Research

It is apparent that in the past, urban research was dominated by state controlled agencies including universities and other related government institutions. Most of the researches done in these institutions were meant to address the specific interests of contracting organisations, thus making them limited in scope and coverage.

There is need to focus more on comparative urban studies involving the experiences of big and small towns in Tanzania. Such studies might reveal common issues of concern cutting across the city centres as well as revealing what should be avoided by the up-coming small towns. At the same time, successful interventions should be studied and replicated.

Urbanisation processes tend to involve more than one actor at any particular time. So, in carrying out such comparative studies on urban centres, a more interdisciplinary (social science) approach involving various stakeholders such as policy makers, planners, demographers, physicians, scientists, politicians, social workers, educationists, psychologists etc.

should be adopted. For instance, in addressing urban poverty, both the government, NGOs, research centres, political parties, religious groups, community based organizations etc need to be involved.

Evidence from other countries - particularly in the North - suggest that urban research is not only confined or concentrated in academic and government institutions, but also involves various stakeholders which Stren (1994: 167) calls policy community. According to Stren, the stakeholders include senior public servants, politicians, private sector entrepreneurs, important voluntary organisations and gender activists. Therefore, it appears that interdisciplinary and inter-sectoral collaboration in urban research is the best approach.

Micro Level Urban Research

In addition to the interdisciplinary, inter-sectoral, comparative studies, there is need to focus on micro-level urban studies. These include case studies that reveal both successful and unsuccessful initiatives being undertaken by the people to deal with urban problems at the household or community level. The Tabata Development Fund is one of the cases worth emulating. Likewise, micro studies will also reveal various strategies used by the urban poor to cope with the urban crisis.

Research on Retrenchment, Migration and Urbanization

Since 1975, the government of Tanzania has implemented three massive retrenchment programmes, which resulted in the loss of public sector employment by a significant number of employees. For instance, in 1975 the government reduced its workforce by 20 percent (Shayo, 1998). Actual records indicate that about 10,000 employees were retrenched. In 1985 about 12,760 employees were retrenched. Since the beginning of 1992, more than 100,000 public sector employees have been dropped from the government payroll as retrenched or redundant employees (Shayo, 1998).

It was assumed that most of these workers would settle in the rural areas and involve themselves in the production of tradable goods (Shayo, 1998). To the contrary, the majority of the retrenched workers have continued to stay in urban areas. Very few retirees have settled in the rural areas. There is need to conduct empirical studies to find out how the retrenched or redundant employees who opted to live in urban areas are coping with the stresses and contradictions of urban life. Most of these retrenched workers used to enjoy certain fringe benefits including housing, transport, and medi-care. After retrenchment fringe benefits were no longer available and they are expected to care for themselves and their families. Most of the retrenched workers from the public sector were given compensatory allowances to allow their smooth transition into the private or informal sectors. It would be interesting therefore, to conduct micro level studies on their investment patterns and type of investments.

Gender and Urbanization

There is a growing recognition of the need to consider gender aspects in socio-economic development. But integrating gender issues is not something 'extra' for practitioners to consider but a regular part of good practice . According to the Habitat Report (1996) integrating gender issues into urban policy planning and management is both an efficiency and equity issue. It promotes equity, and on the other hand it ensures that limited resources are used much more effectively in a manner that benefits every member of the society. Gender considerations in urbanization involves specific actions to address the needs of all genders, interests and priorities. A major prerequisite for this to materialise is to consider designing gender responsive urban policies and programmes.

There is a lot of information on Tanzania that reveals the ways and means through which discrimination against women to access, control and ownership of resources both in urban and rural areas is done (Mukangara, 1994; TGNP, 1993; Koda, 1996; and Shayo, 1998). Surprisingly, there is a dearth of research on gender aspects of urbanization and resource management that is specific to Tanzania (see Mascarenhas this volume). This applies to every aspect of urbanization, including planning, implementation and access to resources. As a consequence, the current planning practice for urban development is gender blind in the sense that gender issues are given a very gestural emphasis, if not all together conspicously absent. Not surprisingly, several gender issues have been underestimated or neglected. These include lack of public toilets for women, poor location of public local markets, poor lighting system and transport network. In fact, most people argue that there is a need to undertake a gender analysis of the urbanization process in order to improve the quality of urban life.

However, gender studies as a discipline is new in Tanzania's teaching curriculum. Therefore, gender training is a prerequisite for improving urban governance. Gender sensitisation programmes ought to be encouraged. Likewise, research is needed to address the existing bureaucratic resistance to gender concerns in the society and particularly in urban areas. Moreover, new institutional arrangements should be created to engender the existing structures and enhance political commitment to these issues.

But it should be noted that training or mainstreaming gender in urban management and governance is a prerequisite and not a panacea for solving urban problems. Integration of gender issues into urban policy planning and management requires that urban planners and policy makers be exposed and knowledgeable on gender analytical tools. There are three levels at which gender analysis is important, namely: macro, meso and micro.

Gender and Urbanisation: Macro Level

Macro issues such as resource mobilisation, access to resource, and unemployment may be associated with or influence the problems associated with urbanisation and poverty with respect to gender. For instance, the massive influx of young boys and girls without adequate skills may have implications for gender differentiated outcomes both in the urban and rural areas. Whereas the young people are increasing the rate of youth unemployment and stretching resources available for urban people, they at the same time cause a shortage of productive labour force in the rural areas.

Gender and Urbanisation: Meso-Level

In Tanzania most studies on gender have paid little attention to women and men issues related to the urbanisation and urban natural resource use, at the meso level (institutional/ enterprise level). Most of the existing accounts on gender issues in almost all social-economic spheres at the meso-level reveal the disadvantaged position of the vulnerable groups, and in particular the urban poor. An initial entry point in engendering the urbanisation process in Tanzania should be the identification of gender issues in the urbanization processes, both at the micro and meso levels. Urban life in Tanzania has changed considerably since the 1980s due to a number of social, economic, and political reforms. The collapse of the Tanzania Housing Bank (THB) and other related institutions has seriously undermined the ability of urban people to build their own houses.

The inability of state enterprises at the meso level to cater for the needs of the rapidly expanding urban population leaves a lot to be desired. For instance, most of the urban poor are engaged in street businesses, and are building unplanned houses in squatter areas. These activities create an additional burden with respect to provision of social services. Policy makers, planners and engineers at the meso level (mostly men) fail to recognise the different and particular needs of the poor (men and women). Hence, there is need to come up with ideas on how to engender policy planning at this level; to accommodate the needs and priorities of the poor people including "*mama lishes*" and people living in river valleys and squatter areas.

Gender and Urbanisation: Micro Level

Although existing policies and the law (at macro level) do not discriminate against women and vulnerable groups at the micro level, most practices are still based on accepted African traditions and customs which discriminate against women . Most of the existing studies on urbanisation at the micro-level (household and community) tend to overlook gender issues. The division of responsibilities, labor, and power within the household, particularly in urban areas, are often underplayed. There is need to identify the gender issues or biases in the urbanisation process at the micro level. For instance, the analysis of gender biases in the urbanisation process at this level is very crucial for the government to réalize its commitment spelt out in the 1998 Draft National Human Settlements Development Policy document namely - intention to involve both men and women in the formulation and execution of workable, broad-based and sound urban policies, right from the village to the national levels.

RESOURCE MOBILISATION FOR URBAN RESEARCH AND DEVELOPMENT

Since the onset of the economic crises in the 1980s, public funding for social and economic development has continued to deteriorate. The adverse economic climate affecting the country has generally caused deterioration in the living standards of the urban households forcing people into developing complex coping strategies. The shortage of funds is also reflected in the funds available for urban research.

For Tanzania, and most Sub-Saharan African countries, urban research support (trained manpower and financial) have all along come from external sources, notably from the North. Stren (1994:3) confirms that overseas agencies play a big role in providing funding for urban research and training for most third world countries.

The concentration of economic reforms and globalization has resulted into the erosion of the overall urban production base (industries and state owned enterprises). To a large extent, the erosion of the economic base will have a direct impact on the financial resources available to support urban research. Worse still, overall financial support from the developed countries to Third World countries is also declining. This implies that the funding of urban research and development is not certain. It is more likely to decline even further.

To counteract the falling rate of urban research funding, there is urgent need for the Tanzania government to consider alternative ways of mobilizing funds for urban research and other development issues. A variety of strategies could be worked out after a thorough assessment of the urban needs, interests and existing opportunities etc. Following this exercise, the government could begin by setting aside funds every year for research and development activities including urbanisation issues. Setting aside of urban research funds similar to the Revolving Housing Loan Fund (RHLF) which was established by the government to assist civil servants to build personal houses, should be encouraged. As a starting point, there is need to create a data bank on the possible sources of urban financing, including local and international.

Industrial Development

Since the mid-1980s the Tanzania economy has been experiencing social and economic crises which forced it to introduce structural adjustment programmes, and in particular industrial reforms. However, the pace of political, economic and social changes has created uncertainties for the industrial base in Tanzania. Most of the industries have collapsed or been privatised.

As a strategy for improving industrialisation in Tanzania there is need to identify strategic areas for industrial development. The primary concern should be directed to the creation of a strategic industrial base for the country that will focus on effective and efficient urban development and management. To-date there are tendencies towards specialization in industrial activities among/across different countries/regions. The ability to compete and sustain industrial bases will highly depend on the strategic advantages a country enjoys *vis-a-vis* other competing countries and its ability to adapt to global changes in industrial and technological development.

As a starting point, there is need to carry out more studies on this problem and come up with strategies on how to operationalise it. Other problems for research are growth trends and patterns of industrialisation, location, size and outputs. Moreover, there is need for more research on the impact of the changing nature of economic and investment policies and rapid technological developments on industrial activities in Tanzania. Furthermore, there is need to focus on the problem of urban infrastructure crisis and its impact on industrialisation.

Urban Finance

The question of financing the urbanisation process remains un-addressed in Tanzania. Indeed, evidence from the contributions in this volume as well as the Draft Human Settlement Development Policy document

(1998) suggests that the question of urban finance has not received adequate attention. Resources available are too limited to meet the demands of the urban population. Available records show that although the Revolving Housing Loan Fund for civil servants was established in 1992, only one sixth of the employees have benefited from it.

In Tanzania so far, there have been very few studies on urban finance. In particular the issue of local and foreign funding has not been streamlined. Therefore, there is need to emphasize research on urban finance, and in particular local finance. Other issues to be studied include problems of financing public and private services in small and big towns, old and new sources of revenue and expenditure patterns between local and external agencies, taxation systems and government plans for financial support and other resources. This is in line with Weiping's (1996: 123) ideas on possible sources for urban finance which include: municipal tax bases, central or regional government, financial markets and capital (venture and other foreign sources for instance the Sida support scheme).

One of the big challenges facing Tanzania in its bid to improve urban management is the question of adequate financing. Urban problems are increasing at a faster rate than the resources available. Nearly all the local and central (urban) administrative tasks still depend on budgetary allocations from the central government. As the state is currently moving away from social service delivery, there is need to search for alternative sources of funds for urban development.

It has become increasingly clear that we cannot improve the quality of our big or small towns without adequate resources, and in particular finance. Moreover, it is recognized that urban development without economic development in the long run will be unsustainable. This link to economic development and improved urbanization draws attention to the problem of urban finance (source, cost and terms), flow of finance and whether urban people can have access to the funds for various socio-economic activities.

Improved Urban Governance, Planning and Management

Like the rest of Sub-Saharan Africa, most of the problems facing urban centres in Tanzania arise from inadequate and inappropriate urban planning and management. Given the current wave of economic reforms (liberalisation and less involvement of the state in social service delivery system) there is need to review the planning approach to urban centres. Indeed, there is urgent need to focus on improved governance and management as outlined in the 1996 Habitat Report.

According to the Habitat Report (1996:160), governance is used in the broadest sense to refer to "the relationship not only between government and state agencies" but also between government and communities and social groups. Undoubtedly, the state has played a dominant role in influencing positively or negatively the pattern of urban growth. There is need for the government to review its management style and promote partnership with civil society and other interested stakeholders.

The Tabata case is a typical example of how community based organisations (CBOs) can be used to promote urban governance. However, there is need to carry on studies that will show the advantages or disadvantages associated with this strategy. It appears that the Tabata CBO gets a lot of support from local and international non-governmental organisations.

Migration and Urbanisation

Rural urban migration has come to prominence as a major demographic phenomenon. It is estimated that over one hundred million migrants in the world are living outside their countries of domicile (White, 1996:164). In Tanzania there is dearth of data to show the exact number of rural-urban migration (Mascarenhas in this volume).

Although the exact factors causing people to move from one point to another (whether urban to rural or rural to urban) are not known, social economic hardships in both urban and rural areas account for why most youths and other people are moving from one point to another. Indeed, one of the main issues of concern is that both rural and urban areas will continue to serve both as the origin and destination of these migrants. Uncontrolled and unplanned migration of people to and from urban and rural areas can itself be the genesis of other social stresses. Therefore, there is need for more in-depth empirical studies on rural-urban migration and urban rural migration to substantiate the above concerns.

It is believed that to-date most of the social problems in the rural areas have been caused by migrants who are moving from urban areas. Other problems like diseases (e.g. HIV/AIDS), social tension and injustice have been reported to be on the increase in the rural areas. On the other hand there is increased tension in urban areas. Most urban people whose origin is in the rural areas are behaving as if they are staying in the rural areas. For instance, in solving their economic problems most of them are engaged in urban farming, dairy, poultry and pig projects. Most of these economic activities are done haphazardly in an unplanned manner. For instance it is common to find some urban people keeping poultry and dairy cattle in residential areas without bothering about the environmental impacts on the neighborhood.

Although urban agriculture and animal husbandry is considered an important component in sustainable development, improperly practiced urban agriculture can result into serious environmental and health hazards. These issues need to be explored. The commitment by the Government of Tanzania to designate special areas where people will be granted legal rights to engage in economic and other agribusiness related activities is highly commendable. At the same time, the need for further research on urban farming and other informal sector activities is more crucial than ever before.

Social Service Delivery

The economic reforms currently taking place in most Sub-Saharan African countries are adversely affecting both the provision and utilisation of various types of social services (health, education) and other economic resources (land, housing etc.) available to most urban people. Since the 1980s, the state has gradually withdrawn or minimised its support for various services. The private sector is increasingly being encouraged to take up the responsibilities of the public sector including social service delivery.

However, the problem of social service delivery lies at the point of intersection between changing economic policies in general (liberalisation, democratisation, bureaucratic reforms) and the socio-economic impacts on the urban population particularly the urban poor. The role of the private sector is worth re-considering. Generally, private sector investments are profit driven. Private investors seek to maximise profits out of their investment whether in education or health or garbage collection. Moreover, they invest where profits are almost certain.

In view of the above observations there is need to consider seriously the issue of social service delivery. Most of the urban poor including civil servants at the lower levels of employment are increasingly finding it difficult to afford different types of services offered by the private sector. For example, primary school education in private schools is as high as Tshs. 450,000 per term. On top of that there are other school activities which demand money. On the average, one needs about T.sh 2 million to send a child to a private secondary school in Tanzania. Obviously, it is more expensive when a child is sent abroad for studies.

Moreover both liquid and solid waste is also an issue of concern to most poor urban people living in squatter areas in big cities such as Dar es

Salaam. Although the government has recently privatised this activity, not adequate efforts have been done to provide services to the urban poor living in squatter ares such as Msimbazi valley squatter areas including Manzese, Sinza areas etc.

Thus, in addressing the multitude of problems related to social service delivery during the era of privatisation, there is need to improve and support the decentralization strategies aimed at developing partnership relations among different stakeholders including, improved CBOs, private companies, local government institutions, NGOs etc. However, the successful implementation of this requires a strong public institution to coordinate and regulate the decentralisation processes and management of the different types of arrangements including expanding job creation both in the formal and informal sectors.

Where do we Begin?

It is clear from the presentation that there is wide scope for further research on urbanisation processes in Tanzania. This chapter has addressed a number of pertinent issues pertaining to urbanisation. It suggests possible areas of concern and possible measures to improve the quality of urban life in the coming years and their implementation. The issues raised are wide but not exhaustive. Much work still remains to be done on what the urban centers would look like in the 21st century. The work presented in this book has raised a few concerns that any re-thinking on the urbanisation processes ought to consider.

As seen in the preceding chapters, urbanisation issues and problems are too complicated for the government and individuals to manage without co-ordination. They encompass social, economic, and political dimensions. Hence, this requires the involvement of all stakeholders. In devolving power away from central authority it is important to think of the following questions: How can we ensure that the interests of the majority of the urban poor are taken care of? What does this mean in terms of new urban governance and management? Given the current wave of economic liberalisation and state withdrawal from social service provision and the emerging powers of certain transnational bodies and institutions (globalisation) how can the marginalised groups (poor men and women in urban and rural areas) best organise themselves to fight for their rights?

From the analysis it appears that political commitment is needed in the prioritization of actions and resources, However, the key issue of identifying the priority area of concern remains a great challenge.

BIBLIOGRAPHY

Cohen, M.A.; Ruble B.A.; Tulchin S.J. and Garland M.A. (1996)
Preparing for the Future: Global Pressures and Local Forces, Washington DC. Woodrow Wilson Centre,

Habitat Report (1996)
An Urbanising World: Global Report on Human Settlement, UN Center for Human Settlement, HABITAT. London, Oxford University Press.

Halfani, M. (1994)
"Urban Research in Eastern Africa: Tanzania, Kenya and Uganda: Towards on Agenda for the 1990s," in, Stren R., *Urban Research in the Developing World,* Vol. 2. Toronto, University of Toronto Press.

Koda, B. (1997)
"The Development of Entrepreneurship in Tanzania: A Gender Perspective," in Ngware S *et al: Gender and Agrarian Change in Tanzania: With a Kenyan Case Study.* Dar es Salaam, DUP (1996) LTD.

Mukangara, F. (1994)
Women and Gender Studies in Tanzania: An Annotated Bibliography. Dar es Salaam, Dar es Salaam University Press.

Shayo, R. (1998)
"Gender, Employment and Structural Adjustment in Urban Tanzania: The Case of the Public Sector Retrenchment Programme 1992–95." Ph.D. Thesis, University of Manchester, Manchester.

Stren, R. (1994)
Urban Research in the Developing World: Vol. 2, University of Toronto, Canada.

Swirling, M. (1997)
Governing African Cities. Johannesburg, Witwatersrand University Press.

TNGP, (1993)
Gender Profile of Tanzania, Dar es Salaam, Dar es Salaam University Press.

Weiping, W.U. (1996)
"Economic Competition and Resource Mobilisation," in Cohen *et al., Urban Future, Global Pressures and Local Forces.* Washington, Woodrow Wilson Press.

White, M.J. (1996)
"Urbanisation and Population Dynamics: City as Villain, Savior or Bystander," in, Cohen *et al., Urban Future, Global Pressures and Local Forces.* Washington, Woodrow Wilson Press.

Author Index

Subject Index

R

S

U

www.ingramcontent.com/pod-product-compliance
Lightning Source LLC
LaVergne TN
LVHW102227300926
849160LV00005B/59
* 9 7 8 9 9 7 6 6 0 3 2 5 5 *